# International Conference on Renewable Energy (ICREN 2018)



E3S Web of Conferences Volume 61 (2018)

Barcelona, Spain
25 - 27 April 2018

**Editor:**

**A. Safdarian**

ISBN: 978-1-5108-7411-4

**Printed from e-media with permission by:**

Curran Associates, Inc.
57 Morehouse Lane
Red Hook, NY 12571

**Some format issues inherent in the e-media version may also appear in this print version.**



Printed by Curran Associates, Inc. (2018)

For additional information, please contact EDP Sciences – Web of Conferences at the address below.

EDP Sciences – Web of Conferences
17, Avenue du Hoggar
Parc d'Activité de Courtabœuf
BP 112
F-91944 Les Ulis Cedex A
France

Phone: +33 (0) 1 69 18 75 75
Fax: +33 (0) 1 69 28 84 91

contact-edps@webofconferences.org

**Additional copies of this publication are available from:**

Curran Associates, Inc.
57 Morehouse Lane
Red Hook, NY 12571 USA
Phone: 845-758-0400
Fax: 845-758-2633
Email: curran@proceedings.com
Web: www.proceedings.com

# TABLE OF CONTENTS

**Conference name:** International Conference on Renewable Energy 2018
ICREN 2018
**Venue:** UPF Barcelona School of Management, Barcelona, Spain
**Dates:** 25-27th April 2018

**Group Picture:**

**Chairman:**

- Prof. Joan Ramon Morante
  Catalonia Institute for Energy Research (IREC)

**Guest Editor:**

- Dr. Amir Safdarian, Sharif University of Technology
  safdarian@sharif.edu

**Reviewers Committee**:

Erik Buchmann, Germany
Jonathan Govaerts, Belgium
Julio Cárabe, Spain
Vladimir Saly, Slovakia
Erdal Bekiroglu, Turkey
Fabio Mottola, Italy
Mark Apperley, New Zealand
Mamadou Lamine Doumbia, Canada
Muhammad Farooq, United Kingdom
Cristina Alonso-Tristán, Spain
Brahim Aïssa, Canada
Igor Konovalov, Germany

John Bögild Hansen, Denmark
Laurent Antoni, France
Mehdi Bagheri, Kazakhstan
Yoshitaka Okada, Japan
Hasimah Abdul Rahman, Malaysia
Essam E. Khalil, Egypt
Paola Delli Veneri, Italy
Amir Safdarian, Iran
Dragan Obradovic, Germany
Alexander Meeder, Germany
PV Aravind, Netherlands
Ahmed Al-Salaymeh, Jordan
Gheorghe-Daniel Andreescu, Romania
Maria Giovanna Minutillo, Italy
Nicola Massarotti, Italy
Arthur W. Weeber, Netherlands
Bruce Nordman, United States
Enrique Romero-Cadaval, Spain
Gianni Bidini, Italy
Subhash, Singhal
Elias Kyriakides, Cyprus
Pierluigi Leone, Italy
Giorgio Graditi, Italy
Yung C. Liang, Singapore
Riccardo Po, Italy
Stefano Ubertini, Italy
Vincenzo Palma, Italy
Stanko Hocevar, Slovenia
Sjoerd Veenstra, Netherlands
Marcello Baricco, Italy
Sachin Kumar , India
Abdelkrim Khireddine, Algeria

# Preface

Renewable Energy is one of the most important challenges that the world must address urgently. Governments and private companies are expanding investments in this field and the research is moving very quickly in several ways. This conference will be the opportunity to gather researchers from different subfields and have a global overview of the latest progress

One of the main targets is to attract the best papers and experts from all around the world. In order to promote excellence, awards will be offered to the outstanding submissions and all participants who would like to publish their full paper will have the opportunity to do it with ICREN 2018.

edp sciences

Web of Conferences

# Statement of Peer review

In submitting conference proceedings to *Web of Conferences*, the editors of the proceedings certify to the Publisher that

1. They adhere to its **Policy on Publishing Integrity** in order to safeguard good scientific practice in publishing.
2. All articles have been subjected to peer review administered by the proceedings editors.
3. Reviews have been conducted by expert referees, who have been requested to provide unbiased and constructive comments aimed, whenever possible, at improving the work.
4. Proceedings editors have taken all reasonable steps to ensure the quality of the materials they publish and their decision to accept or reject a paper for publication has been based only on the merits of the work and the relevance to the journal.

Title, date and place of the conference

INTERNATIONAL CONFERENCE ON
RENEWABLE ENERGY
APRIL 25, 26, 27, 2018
BARCELONA

Proceedings editor(s):

Dr. AMIR SAFDARIAN

Date and editor's signature

08/10/2018

# A simplified model for the dynamic analysis and power generation of a floating offshore wind turbine

*Markus* Lerch[1,*], *Mikel* De-Prada-Gil[1], and *Climent* Molins[2]

[1]Catalonia Institute for Energy Research, Jardins de les Dones de Negre 1, 2a., 08930 Sant Adrià de Besòs, Barcelona, Spain
[2]Universitat Politècnica de Catalunya, Civil and Environmental Engineering Department, Jordi Girona 1-3, 08034 Barcelona, Spain

**Abstract.** This paper presents a simplified model for the dynamic analysis of a floating offshore wind turbine (FOWT), which can be suitable for early feasibility and pre-engineering studies, where the complete system has to modeled in order to predict its behavior and to assess the performance. The model solves the equation of motion in time domain and considers Morison equation for computing the hydrodynamic loads. The aerodynamic loads are included by considering the wind thrust at hub height and the loads from the mooring system have been computed as a non-linear model. A methodology is also presented for calculating the structural properties of the system. The model is tested for two load cases and compared to results obtained with the more complex model FAST. The comparison between the response of the models is satisfactory. The simplified model allows to capture the main motions of the FOWT with an acceptable accuracy. A further feature of the model is to calculate the power generation of the floating wind turbine. The results show that the losses in comparison with a bottom-fixed offshore wind turbine are below 1% or 1.1% according to the load case, which confirms the good performance of the studied FOWT.

## 1 Introduction

Floating substructures for offshore wind turbines are a promising solution that has been under development in recent years. With lower constraints to water depths and soil conditions, floating offshore wind turbines (FOWT) enable to harness the abundant wind resources of deep water areas [1]. As several FOWT concepts have been successfully tested in wave tanks and prototypes have been proven in open seas, floating offshore wind is now reaching a pre-commercial phase where first floating wind turbine arrays are being constructed in European waters [2]. This transition increases the need for comprehensive tools that allow to model the complete system and to predict its behavior as well as to assess the performance. There exist software packages that allow to model wind turbines with a high fidelity and complexity level [3]. However, such programs are either only commercially available or require a comprehensive technical background to be used. The objective of this paper is to present a simpler numerical model that allows capturing the main motions of a FOWT to different met-ocean conditions. In addition, the power generation is obtained considering the specific behavior of the FOWT and compared to a bottom-fixed offshore wind turbine (BOWT). The FOWT studied is the OC3 (Offshore Code Comparison Collaboration) Hywind Spar buoy concept as defined in detail in Ref. [4]. This paper is organized as follows. Section 2, presents the methodologies used in the numerical model. In Section 3, the developed model is validated against the software FAST by performing a dynamic analysis of the FOWT and comparing the results to ones obtained in the research project OC3 [5]. In Section 4, the main conclusions of this paper are summarized.

*Corresponding author: mlerch@irec.cat

## 2 Methodology

### 2.1 Model description

The developed model considers the FOWT as a single rigid body subject to environmental loads and neglects structural deflections by assuming infinite stiffness. This allows for a significant simplification of the model [6]. The dynamic analysis of the FOWT is performed by solving the equation of motion in time domain as follows:

$$(M + A)\,\ddot{x}(t) + B\,\dot{x}(t) + (C)\,x(t) = F_{\text{ext}}(t) \tag{1}$$

where the motion vector *x(t)* represents the displacements in each degree of freedom (DOF). For a FOWT there are typically six rigid-body DOF. Due to the symmetry of the Spar buoy concept and in order to simplify the model only the surge, heave and pitch motions are considered. *B* is the damping and *C* the hydrostatic stiffness. All those before mentioned are 3 x 3 matrices according to the selected DOF. $F_{ext}$ represents the vector of all external forces and moments acting on the FOWT [3]. In order to solve Eq. 1 all the loads and forces have to be identified. The modeling of the external forces is presented in Section 2.2. The methodology for computing the structural properties of the left side of the equation is presented in Section 2.3. Once the equation of motion is completely defined, it is written in the state-space form in order to eliminate the second order differential equations. Afterwards, ode45 function from MATLAB can be used to solve it.

### 2.2 Load description

The forces that act on the FOWT consist of aerodynamic, hydrostatic and hydrodynamic loads as well as the mooring system [3]. Wind and waves are considered as main environmental loads in this study.

#### *2.2.1 Aerodynamics*

The aerodynamic loading on a wind turbine depends mainly on the wind velocity and the rotor characteristics. The wind thrust force acting on the FOWT is given as:

$$F_{\text{wind}} = \frac{1}{2}\,\rho_{\text{a}}\,\pi\,\frac{A_{\text{rotor}}^{\;2}}{4}\,C_{\text{T}}\,v_{\text{rel}}^{\;2}\;\; with \;\; v_{\text{rel}} = v_{\text{wind}} - v_{\text{hub}} \tag{2}$$

where $\rho_a$ represents the air density, $D_{rotor}$ is the diameter of the rotor, $C_T$ the thrust coefficient and $v_{rel}$ the relative wind velocity at hub height. The relative wind velocity is the wind velocity reduced by the hub velocity$v_{hub}$ due to the motions of the substructure. The thrust coefficient is in general a function of the blade tip-speed ratio and the blade pitch angle [7]. This approach has been used in Section 2.4 to calculate the power generation. In regard to the modeling of the structural behavior, a simplified approach was used by considering the dependence of the thrust coefficient only on the wind speed as follows:

$$C_{\text{T}} = \begin{bmatrix} C_{\text{T0}} & if\; v_{\text{rel}} \leq v_{\text{rated}} \\ C_{\text{T0}}\, e^{\,(\text{-a}\,(\text{v}_{\text{rel}}\,-\,\text{v}_{\text{rated}})^{\,\text{b}})} & if\; v_{\text{rel}} > v_{\text{rated}} \end{bmatrix} \tag{3}$$

where $C_{T0}$, a and b are constants. The applied approach allows to maximize the thrust force up to rated wind speed $v_{rated}$ by keeping $C_T$ constant. After rated wind speed, the thrust coefficient is exponentially reduced.

### 2.2.2 Hydrostatics

The hydrostatic loads on the platform refer to the effect of having a submerged body in water. It can be divided into an undisturbed buoyancy force and a restoring term due to the platform movements. The restoring term is the hydrostatic stiffness C of Eq. 1 and its computation is defined in Section 2.3. The buoyancy force is a vertical force directed upwards and according to Archimedes' principle possesses a value equal to the volume of fluid displaced by the body [8] and can be obtained as:

$$F_{\mathrm{buoy}} = \rho_{\mathrm{w}}\, g\, V \quad (4) \qquad\qquad F_{\mathrm{G}} = -\, m_{\mathrm{t}}\, g \quad (5)$$

where $\rho_w$ represents the water density, $g$ the gravitational acceleration and V the displaced submerged volume of the Spar. The force that balances the buoyancy is the weight and is obtained by considering the total mass $m_t$ of the FOWT [4].

### 2.2.3 Hydrodynamics

Morison equation has been applied to calculate the hydrodynamic loads acting on the FOWT. It is one of the widely used methods for slender structures like the Spar and aims to address viscous effects as well as inertial loads by an empirically derived formula [9]. Eq. 6 presents the Morison equation in conjunction with strip theory by dividing the structure in discrete elements of $dz$. The total force is obtained by integrating $dF$ over the length of the Spar [8].

$$dF_{\mathrm{h}} = \frac{1}{2}\rho_{\mathrm{w}}\, C_{\mathrm{d}}\, D\, dz\, |v_{\mathrm{r}}|\, v_{\mathrm{r}} + C_{\mathrm{a}}\, \rho_{\mathrm{w}}\, A(z)\, dz\, a_{\mathrm{r}} + A(z)\, dz\, \rho_{\mathrm{w}}\, a_{\mathrm{W}} \quad (6)$$

The hydrodynamic added mass and viscous-drag coefficients are represented by $C_d$ and $C_a$, respectively. The term $Ddz$ is the frontal area of the strip and $Adz$ is the displaced volume of fluid for the corresponding strip. $v_r$ is the relative velocity between the water particle velocity $v_W$ and the velocity of the body $v_B$ [8]. The equation does not account for the hydrodynamic heave force experienced by the FOWT. The heave force can be approximated by the change of the hydrostatic pressure caused by the variation of wave elevation $\eta$ at the water-plane area $A_{wp}$ as follows [4]:

$$F_{\mathrm{p}} = \rho_{\mathrm{w}}\, g\, \eta\, A_{\mathrm{wp}} \quad (7)$$

### 2.2.4 Mooring system

The mooring loads are modeled by using the quasi-static analysis approach, which considers the offset of the floating structure caused by wave-induced motions and the computation of the non-linear catenary stiffness [10]. The mooring line is taken as a continuous cable with homogeneous properties. However, forces arising from inertia, viscous drag, internal damping, bending and torsion are neglected [11]. The quasi-static model is applied, because it provides a reasonable approximation of the mooring load and a simple calculation methodology compared with a fully dynamic model. As the structure is being displaced, the fairlead position moves at a height h and length l and provokes a resulting horizontal and vertical force at the fairlead from the mooring load. Eq. 8 and 9 are used to obtain the fairlead forces for a fully suspended mooring line [12].

$$l = \frac{X}{w}\left( ln\left( \frac{Z}{X} + \sqrt{1 + \left(\frac{Z}{X}\right)^2} \right) - ln\left( \frac{Z - wL}{X} + \sqrt{1 + \left(\frac{Z - wL}{X}\right)^2} \right) \right) + \frac{XL}{EA} \quad (8)$$

$$h = \frac{X}{w}\left( \sqrt{1 + \left(\frac{Z}{X}\right)^2} - \sqrt{1 + \left(\frac{Z - wL}{X}\right)^2} \right) + \frac{1}{EA}\left( ZL - \frac{wL^2}{2} \right) \quad (9)$$

$X$ represents the horizontal and $Z$ the vertical component of the fairlead force. The unstretched line length is given as $L$ and $w$ represents the weight per unit length of the mooring line in the water. $EA$ is the cross-section axial stiffness. The system of nonlinear equations is solved for a range of possible displacements of the fairlead and by using the solver fsolve from MATLAB. When the vertical force $Z$ is less than the total weight of the cable, then a portion of the mooring line will rest on the seabed and the equations have to be slightly modified as defined in Ref. [12]. The total mooring load on the structure is obtained by considering the fairlead forces of all three mooring lines.

### 2.3 Structural properties

The mass and added mass matrix is obtained as follows [9]:

$$M = \begin{bmatrix} m_t & 0 & m_t\, z_{CoM} \\ 0 & m_t & -m_t\, x_{CoM} \\ 0 & -m_t\, x_{CoM} & I_{yy} \end{bmatrix} \quad A = \begin{bmatrix} \int_{z_{bot}}^{0} \rho_w C_a A(z) dz & 0 & \int_{z_{bot}}^{0} \rho_w C_a A(z) z dz \\ 0 & \frac{2}{3}\rho_w \pi R^3 & 0 \\ \int_{z_{bot}}^{0} \rho_w C_a A(z) z dz & 0 & \int_{z_{bot}}^{0} \rho_w C_a A(z) z^2 dz \end{bmatrix} \tag{10}$$

where $m_t$ is the total mass of the FOWT, $x_{CoM}$ the center of mass and $I_{yy}$ represents the pitch inertia. The added mass is additional mass that the structure appears to have when it is accelerated relative to the surrounding water. Strip theory is used to calculate the added mass for each DOF using constant added mass coefficients of two dimensional sections and integrating over the length [13]. $C_a$ represents the added mass coefficient and $A(z)$ the cross-sectional area of the Spar structure. Ref. [4] recommends to add linear damping to capture correctly the response of the OC3-Hywind concept to hydrodynamic loads. The damping for surge and heave are $1.0e^5$ N s/m and $1.3e^5$ N s/m, respectively. The hydrostatic stiffness represents the restoring term as effect of the substructure movements in the water in heave and pitch direction and is computed as follows, where $z_{CoB}$ is the center of buoyancy [14].

$$C = \begin{bmatrix} 0 & 0 & 0 \\ 0 & \rho_w g A_{wp} & 0 \\ 0 & 0 & \rho_w g I_{wp} + \rho_w g V z_{CoB} - m_t g z_{CoM} \end{bmatrix} \tag{11}$$

### 2.4 Power generation

The power generated by the FOWT can be calculated by Eq. 12 taking into account the rotor swept area $A_{rotor}$, the power coefficient $C_p$ and the wind speed $v_{wind}$ at hub height. The power coefficient depends on the blade tip-speed ratio $\lambda$ and the blade pitch angle $\beta$ [7].

$$P_{FOWT} = \frac{1}{2}\rho_a\, A_{rotor}\, C_p(\lambda, \beta)\, (v_{wind}\, cos(\theta))^3 \tag{12}$$

For the case of a FOWT two considerations have been included. The first is that the motions of the FOWT provoke an additional mean platform tilt angle. This causes the rotor to be slightly titled against the inflow wind velocity. This effect is taken into account in the power calculation by reducing the wind velocity by the pitch angle $\theta$ of the structure [3]. The second consideration is that the model takes into account the relative wind velocity in the pitch computation of the FOWT as outlined in Section 2.2.1.

## 3 Model validation

### 3.1 Static sizing

In this section, the obtained static properties of the FOWT are presented and compared with the ones computed by FAST in the OC3 project [15].

$$\begin{bmatrix} 8.07e^6 kg & 0 & -6.29e^8 \text{kg m} \\ 0 & 8.07e^6 kg & 1.12e^5 \text{kg m} \\ -6.29e^8 \text{kg m} & 1.12e^5 \text{kg m} & 6.80e^{10} \text{kg m}^2 \end{bmatrix} \quad \begin{bmatrix} 8.07e^6 kg & 0 & -6.29e^8 \text{kg m} \\ 0 & 8.07e^6 kg & 1.12e^5 \text{kg m} \\ -6.29e^8 \text{kg m} & 1.12e^5 \text{kg m} & 6.80e^{10} \text{kg m}^2 \end{bmatrix}$$

Mass computed by FOWAT Mass computed by FAST

The mass matrix calculated by the developed model agrees well with the one obtained by FAST [4, 16]. The obtained added mass matrix is presented next and compared to the results from FAST for zero frequency [4].

$$\begin{bmatrix} 7.98e^6 kg & 0 & -4.94e^8 \text{kg m} \\ 0 & 2.23e^4 kg & 0 \\ -4.94e^8 \text{kg m} & 0 & 3.97e^{10} \text{kg m}^2 \end{bmatrix} \quad \begin{bmatrix} \approx 8.00e^6 kg & 0 & \approx -4.90e^8 \text{kg m} \\ 0 & \approx 2.00e^4 kg & 0 \\ \approx -4.90e^8 \text{kg m} & 0 & \approx 3.90e^{10} \text{kg m}^2 \end{bmatrix}$$

Added mass computed by FOWAT Added mass by FAST approximated

The mooring stiffness matrix obtained by FOWAT is shown next. The accuracy of the developed model is quite high for the mooring stiffness calculation in comparison to FAST.

$$\begin{bmatrix} 4.12e^4 \text{N/m} & 0 & -2.82e^6 \text{N/rad} \\ 0 & 1.19e^4 \text{N/m} & 0 \\ -2.82e^6 \text{N/m} & 0 & 3.11e^8 \text{N m/rad} \end{bmatrix} \quad \begin{bmatrix} 4.12e^4 \text{N/m} & 0 & -2.82e^6 \text{N/rad} \\ 0 & 1.19e^4 \text{N/m} & 0 \\ -2.82e^6 \text{N/m} & 0 & 3.11e^8 \text{N m/rad} \end{bmatrix}$$

Mooring stiffness computed by FOWAT Mooring stiffness computed FAST

The obtained hydrostatic matrix is shown next. The hydrostatic stiffness in pitch considers only the effect of the hydrostatic pressure as defined in the OC3 report Ref. [4]. The results of both models are in good agreement.

$$\begin{bmatrix} 0 & 0 & 0 \\ 0 & 3.34e^5 \text{N/m} & 0 \\ 0 & 0 & -5.01e^9 \text{N m/rad} \end{bmatrix} \quad \begin{bmatrix} 0 & 0 & 0 \\ 0 & 3.33e^5 \text{N/m} & 0 \\ 0 & 0 & -4.99e^9 \text{N m/rad} \end{bmatrix}$$

Hydrostatic stiffness computed by FOWAT Hydrostatic stiffness computed by FAST

Based on the previously presented static matrices, the natural frequencies and periods of the FOWT are computed and presented in Table 1.

**Table 1:** Natural frequencies and periods

| | Surge Frequency | Heave Frequency | Pitch Frequency | Surge Period | Heave Period | Pitch Period |
|---|---|---|---|---|---|---|
| FOWAT | 0.008 Hz | 0.033 Hz | 0.033 Hz | 124.003 s | 30.002 s | 30.776 s |
| FAST | 0.008 Hz | 0.032 Hz | 0.034 Hz | 125.000 s | 31.250 s | 29.412 s |

The surge frequency obtained from the FOWAT model matches the value calculated in the OC3 report in Ref. [5]. The frequency in heave is slightly higher than the reference one and the pitch is slightly lower. However, the differences are smaller than 3% and the accuracy of the developed model is seen to be sufficient for the purpose of this study.

### 3.2 Dynamic analysis

In this section, the dynamic response of the FOWT to two load cases (LC) is computed. The LC are based on the OC3 Phase IV study (Ref. [15]) where different modeling codes have been compared.

#### *3.2.1 Load case 1*

The first LC is used to analyze the behavior of the FOWT based on the excitation by a steady wind force of 8m/s and regular airy waves of 6m height and 10s period. The time response for the non-transient part is shown in Fig. 1.

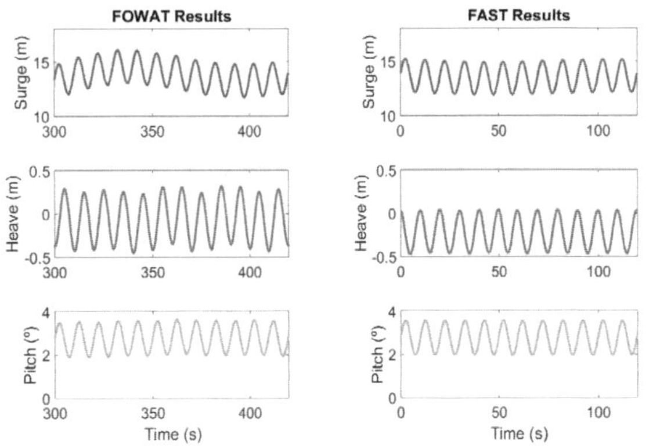


**Fig. 1.** Non-transient response

| | FOWAT | FAST |
|---|---|---|
| Surge | 13.68 m | 13.54 m |
| Heave | -0.07 m | -0.22 m |
| Pitch | 2.74 ° | 2.75 ° |

**Table 2:** Mean displacements

It can be observed that the system oscillates around the equilibrium position and with the wave frequency in all degrees of freedom. The oscillation with the natural frequencies is also visible. Besides that, it can be seen that the wind force generates an offset in both the surge and pitch DOF, which causes the equilibrium point to be different than zero for these two. The average values for the non-transient part obtained with the developed model and FAST are presented in Table 2. The values are close to the ones obtained with FAST, which allows to conclude that the aerodynamic effect is correctly captured by the model.

#### *3.2.2 Load case 2*

LC 2 is used to study the effect of irregular waves and turbulent wind based. JONSWAP spectrum is used to create the irregular wave profile with a significant wave height of 6m and a peak-spectral wave period of 10s. The turbulent wind, based on the Kaimal spectrum, has a mean wind speed equal to the rated speed of 11.4m/s and a turbulence intensity of 0.14. Since the irregular wave profile is a superposition of waves with different frequencies, the response of the FOWT is shown as statistical parameters in Table 3.

**Table 3:** Response comparison between FOWAT and FAST

| | | Wind | Wave | Surge | Heave | Pitch |
|---|---|---|---|---|---|---|
| Minimum | FOWAT | 6.28 m/s | -4.54 m | 14.67 m | -0.83 m | 0.18 ° |
| | FAST | 6.60 m/s | -5.84 m | 11.38 m | -1.07 m | 1.33 ° |
| Mean | FOWAT | 11.11 m/s | 0.01 m | 23.79 m | -0.21 m | 4.74 ° |
| | FAST | 11.43 m/s | 0.01 m | 21.19 m | -0.47 m | 4.25 ° |
| Maximum | FOWAT | 16.16 m/s | 4.73 m | 31.78 m | 0.23 m | 7.12 ° |
| | FAST | 17.37 m/s | 4.73 m | 31.13 m | 0.11 m | 6.26 ° |
| Standard | FOWAT | 1.46 m/s | 1.36 m | 3.84 m | 0.17 m | 1.16 ° |
| Deviation | FAST | 1.96 m/s | 1.49 m | 4.09 m | 0.22 m | 0.84 ° |

For this LC the range of motions show a good agreement with the mean values calculated by FAST. A slight over- or underestimation is observable for some of the minimum and maximum values, which could be due to the statistical estimation of the loads.

### 3.3 Power generation performance

The power of the FOWT has been calculated for a range of wind velocities and wave heights to simulate its specific power curve. The environmental conditions considered are regular waves and a steady wind velocity. A power curve has been computed for each of the wave heights as illustrated in Fig. 2.

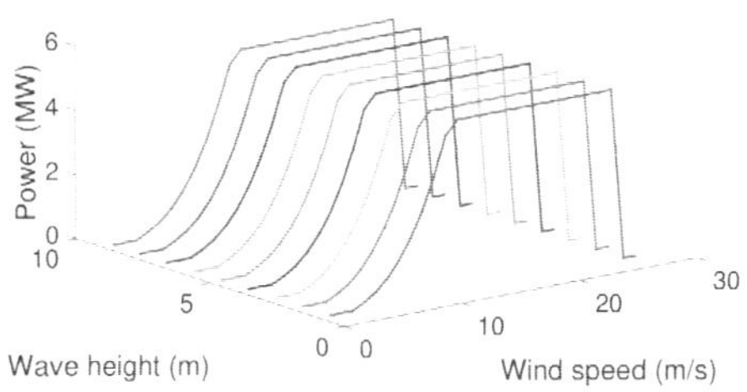


**Fig. 2.** Power curves for different waves

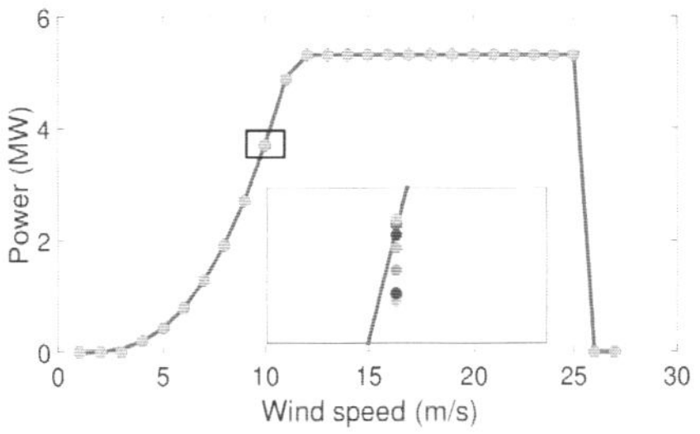


**Fig. 3.** Power curve comparison for LC 1

The power curves include the specific consideration for a FOWT as explained in Section 2.4 and the cut-in and cut-out wind speed limits of the wind turbine. It is observable that the FOWT behaves very stable to different wave heights, since the power curves are very similar. This behavior is very characteristic for a Spar-type floating substructure, because the deep draft and large inertia result in low heave and pitch motions in operating conditions [17]. Fig. 3 shows the comparison between the original power curve of the NREL BOWT and the ones obtained by the FOWT. The blue line represents the power curve of the BOWT and the one of the FOWT is marked by the dots. The subplot indicates the difference according to the wave heights. The power curve of the FOWT is nearly identical to the one obtained by the BOWT. Even the largest difference between the power curve of the BOWT and the most extreme wave is only smaller than 1%. The power curve is computed next considering an environment with irregular waves and a turbulent wind velocity (Fig. 4), which represents a more realistic offshore scenario.

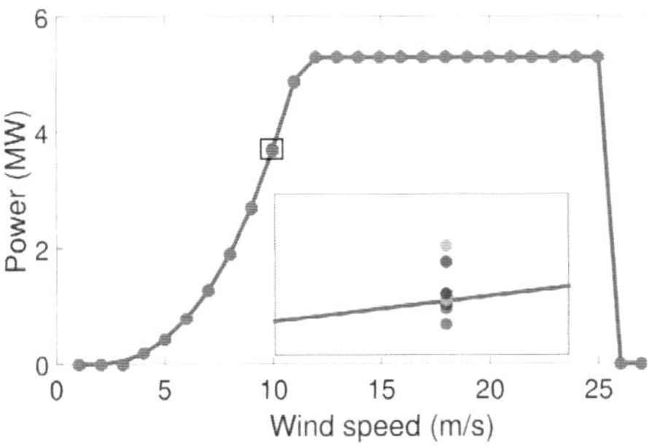


**Fig. 4.** Power curve comparison for LC 2

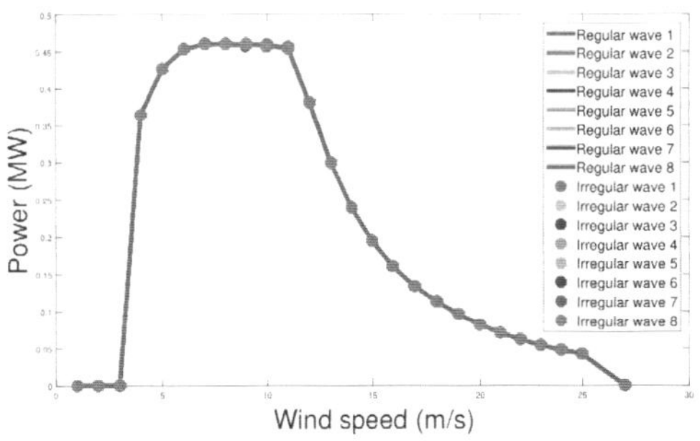


**Fig. 5.** Power coefficient comparison

It is observable that the power curves for the FOWT follow the power curve obtained by the BOWT. The largest difference between the power curve of the BOWT and the most extreme wave is about 1.1% and is, therefore, only slightly higher than compared to the regular wave and steady wind LC. Finally, Fig. 5 shows the power coefficient obtained for the FOWT and confirms that there is a non-significant difference between the LC and the waves. It can be concluded that the wind and wave loads have a non-significant effect on the power production performance of the OC3 Spar buoy FOWT and that the power generation is comparable with a BOWT. This performance has also been demonstrated in experimental tests of the Hywind and WindFloat prototypes in real offshore conditions [3].

## 4 Conclusions

In this paper, a simplified model has been presented for the static and dynamic analysis of a Spar buoy FOWT. It was investigated the effect of different LC on the platform motions and it was studied the potential difference in the power generation of a FOWT and a BOWT. The model was built by using MATLAB and the system response was evaluated for the surge, heave and pitch motions. The results have been compared with FAST, which is a well-known complex tool to model and simulate wind turbines. An overall good agreement has been found in the comparison of the structural properties computed by both models. Furthermore, the main motions and system's dynamics could be captured by the simpler model with an acceptable accuracy. In addition, the power generated by the FOWT has been computed for an environment with regular waves and steady waves as well as a LC consisting of turbulent wind and irregular waves. It was found that even for the most extreme conditions the power loss is less than 1% or 1.1%, respectively the load case studied. This demonstrates the high power performance of the OC3-Hywind Spar buoy under different met-ocean conditions and also coincides with the experiences of prototype tests of FOWT demonstration projects. Further research is suggested to include the energy generation in the model, which can be used for a LCOE estimation. Besides that, the model could be applied to different offshore sites in order to investigate the performance of the FOWT in realistic met-ocean conditions.

## Acknowledgements

This work was supported in part by the European Union Horizon 2020 programme under the grant agreement H2020-LCE-2014-1-640741.

## References

[1] R. James, M.C. Ros, The Carbon Trust (2015)

[2] WindEurope, *Floating Offshore Wind Vision Statement* (WindEurope, 2017)

[3] J. Cruz, M. Atcheson, *Floating Offshore Wind Energy: The Next Generation of Wind Energy* (Springer, 2016)

[4] J. Jonkman, Tech. rep., National Renewable Energy Laboratory (NREL), Golden, CO. (2010)

[5] J. Jonkman, T. Larsen, A. Hansen, T. Nygaard, K. Maus, M. Karimirad, Z. Gao, T. Moan, I. Fylling, Tech. rep., National Renewable Energy Laboratory (NREL), Golden, CO. (2010)

[6] C. Curfs, Master's thesis, Delft University of Technology (2015)

[7] D.B. Fernando, d.B. Hernán, J.M. Ricardo, *Wind Turbine Control Systems: Principles, Modelling and Gain Scheduling Design* (Springer, 2006)

[8] J.M. Jonkman, *Dynamics modeling and loads analysis of an offshore floating wind turbine* (National Renewable Energy Laboratory (NREL), Golden, CO (United States), 2007)

[9] M. Karimirad, *Offshore energy structures: for wind power, wave energy and hybrid marine platforms* (Springer, 2014)

[10] S. Chakrabarti, *Handbook of Offshore Engineering (2-volume set)* (Elsevier, 2005)

[11] M. Masciola, J. Jonkman, A. Robertson et al., *Implementation of a multisegmented, quasi-static cable model*, in *The Twenty-third International Offshore and Polar Engineering Conference* (International Society of Offshore and Polar Engineers, 2013)

[12] M. Masciola, NREL. Available online: https://nwtc. nrel. gov/system/files/MAP_v0 (2013)

[13] M. Karimirad, T. Moan, Marine Structures **27**, 45 (2012)

[14] E.E. Bachynski, T. Moan, Marine Structures **29**, 89 (2012)

[15] J. Jonkman, W. Musial, Tech. rep., National Renewable Energy Laboratory (NREL), Golden, CO (United States) (2010)

[16] J. Jonkman, *Inertial moments of oc3-hywind components*, https://wind.nrel.gov/forum/wind/viewtopic.php?t=748 (2016), accessed: 2018-03-07

[17] S. Nallayarasu, N.S. Kumar, Ships and Offshore Structures **12**, 19 (2017)

# An offshore wind farm energy injection mastering using aerodynamic and kinetic control strategies.

*Djamel* Ikni[1], *Ahmed Ousmane* Bagre[2*], *Mamadou Baïlo* Camara[1], *and Brayima* Dakyo[1].

[1] Group of Research in Electrotechnics and Automatic of Le Havre *(GREAH)*, University of Le Havre, France
[2] Laboratory of Renewable Energy and Energy Efficiency, International Institute for Water and Environmental Engineering -2iE Ouagadougou, Burkina Faso

**Abstract:** The injection of wind farm production into a grid, needs optimal strategies for energy transfer management. Usually, the power produced by the wind farms does not fulfil all the grid code requirements. The main problem is generally based on the way to reduce the impacts of power production fluctuations on the grid voltage and its frequency. To solve this problem, some authors suggest the use of an interface such as energy storage devices in order to compensate the wind power fluctuations. In fact, the storage devices installed between the wind farm and the grid can improve the power quality in terms of stability but in other hand the size and the cost of the system can be increased. In this paper, two solutions have been proposed in case the power quality produced by the wind farm is out of the grid code requirements. The improvement of the energy quality of an offshore wind farm without storage and connected to the grid is discussed. The proposed solution is to operate the wind turbines with a reserve of power. To distribute this reserve equitably among wind turbines, a proportional distribution algorithms has been developed. The results obtained show clearly the effectiveness of the strategy.

**Keywords**: Offshore wind farm, power fluctuation, wake effects, power management, grid code requirement.

## 1. Introduction

The European offshore wind power contribution in 2020 is estimated at 40% from the overall electric production from renewable sources. For the year 2020, the part of wind energy in the total energy production in Europe is predicted to be around 20% [1], [2]. In France, the part of wind energy in year 2017 estimated at 4,1% of the overall energy consumed [3] and for the year the target is to increase the renewable energy from 9.6% in 2005 to 23% in 2020 [4].

The rapid growth of wind energy worry grid managers because of its fluctuation behaviour in terms of power production. Moreover, these networks can achieve very rapidly significant wind energy penetration levels , as it's the case in three European countries: Denmark with 24.4%; Germany with 14.8% and Spain with 14.1% of the total capacity of the grid [5]. The estimated limit beyond which the fluctuations of wind power become very complex to manage, is between 20% and 30% [6]. In this context, all grid managers wish to limit the impacts of the power fluctuations for a better grid stability. These limits are generally summarized in grid codes [7]-[8]. Therefore, it is necessary to propose appropriate solutions in spite of the rapid growth of wind energy with the respect of power quality.

The main objective of this study is to evaluate the capacity of the wind farm to inject properly its production by using pitch angle control and inner inertial reserve. The first section is related to wind distribution model development in an offshore wind farm. The second section proposes some solutions for power quality improvement in the case the energy parameters are not compatible to the limits imposed in Germany's grid code used in this paper as a reference. The studied offshore wind farm system configuration is presented in Fig.1.

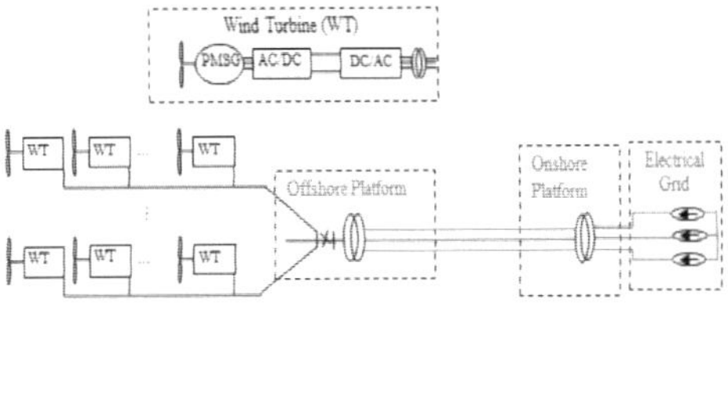


* Corresponding author: ahmed.bagre@2ie-edu.org

**Fig. 1.** Topology of the studied offshore wind farm system .

## 2. Wind distribution modelling for wind farm

A critical point in the wind farm study is the wind distribution model. Many models have been developed in order to get their real behaviour, but unfortunately they need several parameters to be computed [9]-[10]. The proposed model takes into account the inequality distribution of the wind speed in the wind farm. This distribution is due to the strong interactions between wind speed and turbines positioned in first sectors in the direction of wind speed. These interactions cause wake effects which are considered in the model proposed. In [11] and [12], a simplified representation of wake effect is presented where, the wind speed diminished about 2-4% after each row/column of the farm. The same approach is used in the model proposed for wind distribution in wind farm. To reflect also the wind speed delay in the spread between the wind turbines, a delay function is added in wind speed model.

In [13] and [14], the authors have shown that when the wind direction is perpendicular to the lines of the wind farm, the power produced from the wind farm presents more fluctuations. The assumption made in this paper is that the wind speed is the same for all wind turbines positioned in the same line. The wind speed model for (i) line is presented in (1), where v1 is the wind speed for the first line (i=1), dti is the delay function, and dsi presents the wake effect.

These parameters depend on the wind farm configuration. The configuration is made to maximize the production and minimize the turbulence effect between downstream and upstream turbines. For this raison, an optimum distance between the wind turbines is necessary as described in [9] and [15]. This distance is roughly 7 times of wind turbine's diameter. The figure 2 shows the configuration of the wind farm studied with a power rated of 300MW. This power is produced by 60 Permanent Magnets Synchronous Generators (PMSG) linked to turbines. The generators are distributed in 10 lines, and each line includes 6 wind turbines which are in the same sector and submitted to the similar wind speed.

$$v_i = v_1(t - dt_i)(1 - ds_i) \tag{1}$$

The power produced by the wind farm is the sum of the contributions of all the turbines. These contributions are related to wind speeds profiles applied to each turbine. For this raison, it is necessary to calculate dti and dsi parameters as is given in (2) and (3), respectively.

$$dt_i = \frac{7\,D}{v_{1mean}} i \tag{2}$$

$$ds_i = b_i \tag{3}$$

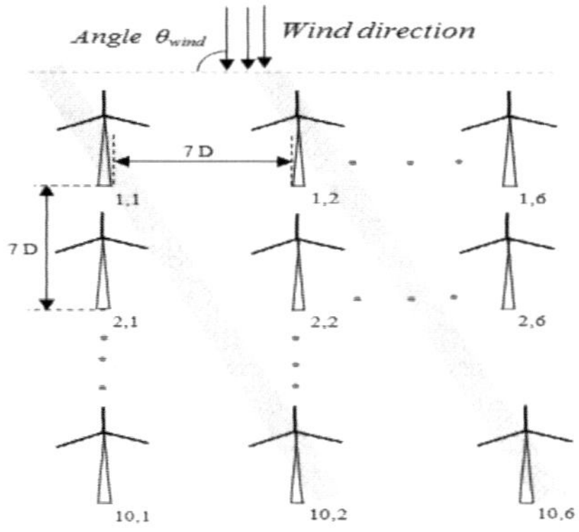


**Fig. 2.** Geometry of wind turbines positions in offshore wind farm.

Where, $v_{1mean}$ is the average value of the wind speed for first line (i=1), D presents the wind turbine diameter, and b is between 2% and 4%. b has been fixed at 2% in this paper. Using (2) and (3) in (1), the wind speed profile for each wind turbine can be estimated as presented in equation (4).

$$v_i = v_1(t - \frac{7D}{v_{1mean}} i)(1 - b_i) \tag{4}$$

The total power produced by the wind farm is given in (5), where $Pi$ presents the power captured by the turbines of line $i$. $Pi$ is as presented in (6), where, $\rho$ is the air density (1.225kg/m$^3$); R is the radius of the rotor blade in (m) and $Cpmax$ is the maximum power coefficient.

$$P_p = 6\sum_{i=1}^{10} P_i \tag{5}$$

$$P_i = \begin{cases} \frac{1}{2}\rho\pi R^2 C_{pmax} v_i^3 & v_i \leq 11{,}5\, m/s \\ 5MW & v_i > 11{,}5\, m/s \end{cases} \quad (6)$$

### 2.1. Wind speed model evaluation

To evaluate the wind speed model, the wind speed profile for the first line turbines presented in Fig. 3 is used. The average value of this profile is around 10.05 m/s with the fluctuations of ± 15%. The wind speed model evaluation approach is based on a comparison between the wind farm power fluctuations with those extracted in real applications described in [16] and [17]. The method used to quantify power fluctuations, is based on the spectrum analysis of powers plotted in Fig. 4. The obtained spectres from computed power (Pp) and reference ones (Pref) are illustrated in Fig. 5. The comparison criterion is based on an analysis of Fluctuation Harmonic Content (FHC) parameter. The FHC is equal to the Normalized Standard Deviation (NSD) of the power in the time domain described in [14]. FHC parameter can be estimated using (7), and where P0 is the produced power average value for 10 minutes, F presents the frequency interval. In this equation P(f) correspond to spectrum analysis of the power using Fast Fourier Transform (FFT).

$$FHC(F) = \frac{\sqrt{\sum_{f\in\{F\}}\left({P(f)}/{\sqrt{2}}\right)^2}}{P_0} \quad (7)$$

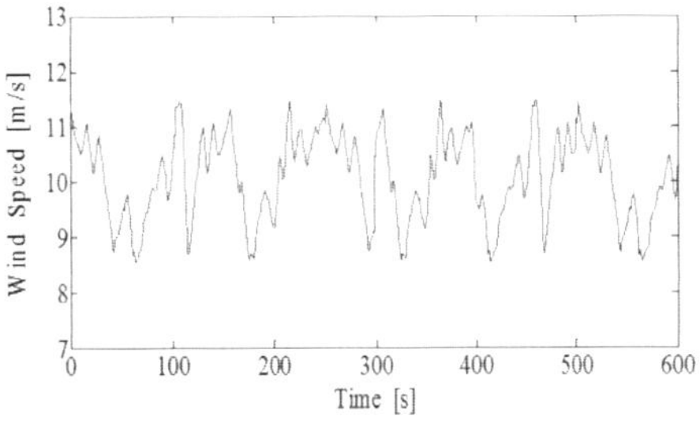


**Fig. 3.** Wind speed profile $v_l$ for first line turbines.

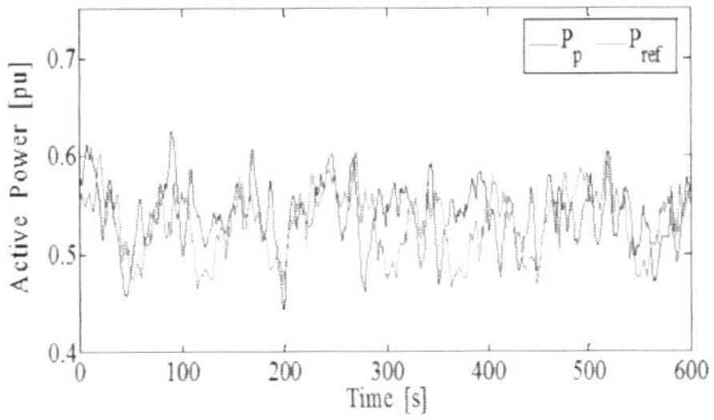


**Fig. 4.** Computed power compared $P_p$ to measured power in real application $P_{ref}$.

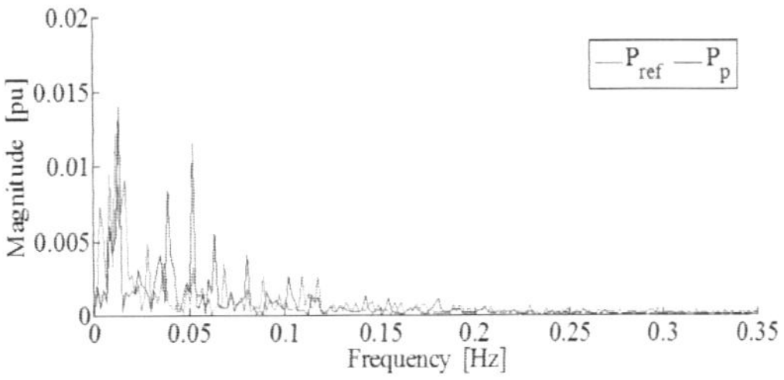


**Fig. 5.** Spectrum analysis of computed power and reference power.

The spectrum of each power is divided into three frequency domain:

- First domain is based on low frequencies which are lower than 0.01Hz.
- The second concerns the medium frequencies which are between 0.01 to 0.3Hz.
- The last zone corresponds to high frequencies which are above 0.3Hz.

FHC parameters estimated for Pref and Pp are summarized in (Table I).

**Table 1.** FHC parameters estimated for $P_{ref}$ and $P_p$

| Frequency | FHC of $P_{ref}$ | FHC of $P_p$ |
|---|---|---|
| f< 0.01Hz | 2.145 | 2.18 |
| f ϵ [0.01- 0.3Hz] | 2.37 | 2.4 |
| f > 0.3Hz | 0.035 | 0.02 |
| Full region | 4.32 | 4.44 |

The estimated values of FHC in each zone are the same. The FHC values corresponding to high frequencies domain are negligible for two powers. For medium frequencies domain, FHC is about 2.37% for Pref, and 2.4% for Pp which presents a difference of 0.03%. In the case of low frequencies domain, the values of FHC are 2.145% for Pref, and 2.18% for Pp which correspond to a difference of 0.035%. These differences are very small and allows us to validate the model of the wind speed distribution proposed for wind farm application.

## 3. Energy quality compatibility problems between wind farm production and grid code exigencies

To inject energy from wind farm into the grid, the power from the wind farm must comply with the grid code requirements. In this paper, the requirements are focused on energy quality and power gradient $(dP/dt)_{\text{limit}}$. These limits are fixed through the grid codes which are not the same for all the countries. For example, the power gradient limits fixed by countries such as Germany, Denmark and Nordic are shown in (Table 2).

**Table 2.** Power gradient limits extracted in [7]-[8]

| Grid code | Power gradient (dP/dt)limit |
|---|---|
| E-on, Germany | 10% of $P_{nom}$ per minute |
| Eltra, Denmark | 10-100% of $P_{nom}$ per minute |
| Nordic Grid Code | 10% of $P_{nom}$ per minute |

To verify whether the power from the wind farm is compatible to the grid code requirement, the ratio given in (8) is used, where r presents 10% of rated power (0.5MW/s) and $\tau_r$ is the sampling time in [s].

$$r = \frac{p_p(t+\tau_r)-p_p(t)}{\tau_r} \tag{8}$$

This equation is used to estimate the required power Pr profile for the grid using wind farm production Pp. The resulting profile which corresponds to the grid code requirement is presented in Fig. 6.

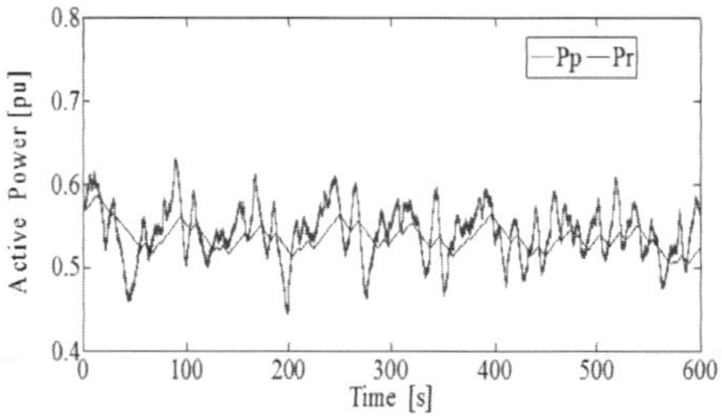


**Fig. 6.** Wind farm production $P_p$ and required power profile for the grid $P_r$.

The figure 6 shows that, the wind farm production Pp has more fluctuations compared to the required power Pr of the grid. The difference between Pp and Pr must be compensated by control strategies, which is the main contribution of this paper. The strategies proposed are based on the downgrading operations of wind farm called in this paper as "power's reserve". These strategies enable to smooth the wind farm production in order to make at any time the total production becomes compatible to what is required for injection into the grid.

## 4. Wind farm power's reserve

The power's reserve in the wind farm can be obtained by reducing its power compared to its maximum capacity. Wind turbine operating in below its maximum power enables to create the power's reserve by action on the wind turbine speed or pitch angle. These actions allow obtaining a power's profile which satisfies the grid code requirements. Two control methods are possible to reduce the power from wind turbine through an action on power's coefficient $C_p(\lambda, \beta)$ given in equation (9) [18].

$$\begin{cases} C_p(\lambda, \beta) = 0{,}73\left(\frac{151}{\lambda_i} - 0{,}58\,\beta - 0{,}02\,\beta^{2.14} - 13{,}2\right)\exp\left(\frac{18{,}4}{\lambda_i}\right) \\ \lambda_i = \frac{1}{\frac{1}{\lambda - 0{,}02\beta} - \frac{0{,}003}{\beta^3 - 1}} \end{cases} \tag{9}$$

The first method is based on the acceleration and the deceleration of the wind turbine according to the profile needed. This technique enables to store energy in kinetic form. The second technique is related to pitch angle control which enables also to store energy in aerodynamic from.

### 4.1. Kinetic reserve principle

For slower speeds than nominal ones, the wind turbine control optimizes the power by adjusting rotational speed. To describe the kinetic reserve principle, the pitch angle is assumed constant, so that, the power's coefficient $C_p$ is the only parameter according to the rotational speed as illustrated in Fig. 7.

The kinetic reserve is obtained by a difference between the power estimated at Ωtr-opt and the one computed at Ωtr-nominal. This operation is getting by action on the electromagnetic torque of the generator [19].The figure 8 shows the kinetic reserve (ΔPIR-max) according to wind speed, which is limited to its nominal value of 11.5m/s. The maximum value of the kinetic reserve for each wind turbine can be estimated using (10). This equation shows that, the kinetic reserve is not proportional to the wind speed.

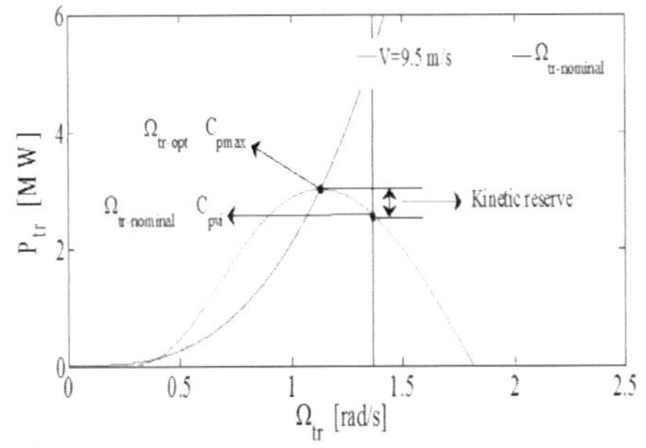


**Fig. 7.** Principle illustration of kinetic reserve.

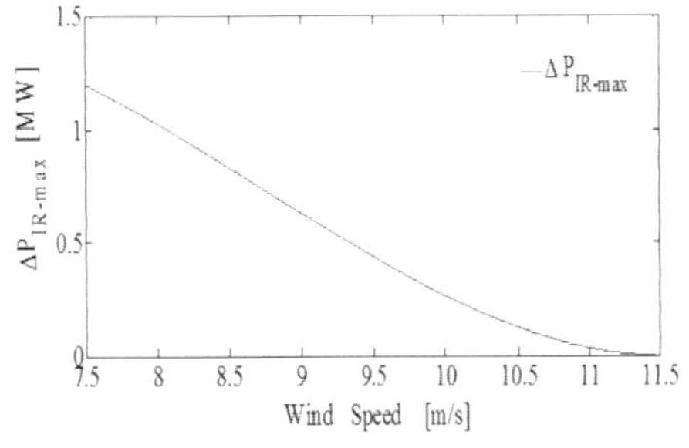


**Fig. 8.** Maximum kinetic reserve according to wind speed.

$$\begin{cases} \Delta P_{IR-max} = P_{tr}\left(C_{Pv_i}, \Omega_{tr-nominal}\right) - P_{tr}\left(C_{p\,max}, \Omega_{tr-opt}\right) \\ C_{pvi} = f(\beta = 0, \lambda_v), \lambda_v = \frac{R \cdot \Omega_{tr-nominal}}{V} \end{cases} \tag{10}$$

### 4.2. Aerodynamic reserve principle

The pitch angle β control enables to reduce the power extracted from wind turbine by action on blades orientation. This solution can be used for all operation zone of the wind turbine as illustrated in Fig. 9 [20].

In normal operation, the pitch angle control limits the power captured by wind turbine at its nominal value. The increase of the Pitch angle reduces the lift of the blade which reduces the wind turbine torque and corresponding power as illustrated in Fig. 9. Whatever the wind speed, it is possible to act on the pitch angle to change the extracted power. The difference between available power in the wind and the power captured by the turbine after pitch angle action presents an aerodynamic reserve.

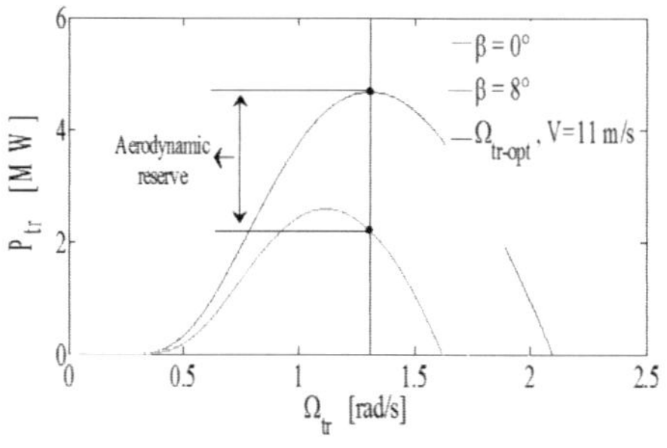


**Fig. 9.** Principle illustration of aerodynamic reserve.

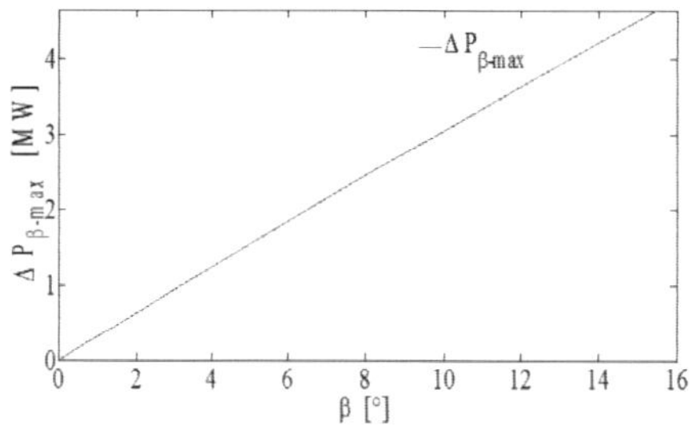


**Fig. 10.** Aerodynamic reserve according to pitch angle $\beta$, for a wind speed of 11m /s.

The figure 10 shows that the aerodynamic reserve due to pitch angle variation can be significant. If necessary, this technique allows obtaining a power reserve equal to the maximum power that the turbine can extract from the wind. The operating point corresponds to the wind turbine feathering position. Aerodynamic reserve can be estimated as given in (11), where Cpmax is the maximum power coefficient without pitch angle, $\Omega_{tr\text{-}opt\text{-}vi}$ presents optimal rotational speed corresponding to the wind speed of $v_i$, and $C_{p\beta}$ presents power coefficient corresponding to a pitch angle value. In this technique, the generator's rotational speed is fixed to its optimum value through the classical Maximum Power Point Tracking (*MPPT*) method.

$$\Delta P_{\beta-max} = P_{tr}\left(C_{p\,max}, \Omega_{tr-opt}\right) - P_{tr}\left(C_{p\beta}, \Omega_{tr-opt-vi}\right) \tag{11}$$

### 4.3. Required power profile estimation for the grid using "grid code" requirements

The algorithm used to determine the required power reserve is given in Fig. 11. The aim of this algorithm is to calculate graphically the value of α in order to get the power authorized for injection into the grid $P_{r\text{-}ref}$ becomes less than the power available in the wind farm $P_{p\text{-}avail}$. This condition is obtained by adjusting the value of $\alpha$ as presented in (13).

$$r = \frac{P'_{p-avail}(t+\tau_r) - P'_{p-avail}(t)}{\tau_r} \tag{12}$$

$$\begin{cases} P_{r-ref} - P_{p-avail} \leq 0 \\ P'_{p-avail} = \alpha P_{p-avail} \end{cases} \tag{13}$$

The relationship between Rp, the power reserve and α coefficient is given in (14). The parameters estimated of the profile used to simulate the system are: Rp =17.5% and α is 82.5%. To smooth the available power Pp-avail of 319 MW, a power reserve of 17.5% necessary is nearly 56 MW.

$$R_p = 1 - \alpha \tag{14}$$

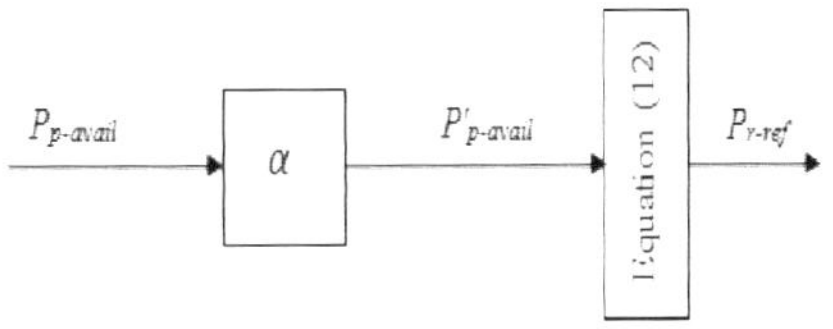


**Fig. 11.** Power required profile estimation for the grid.

### 4.4. Aerodynamic reserve selection compared to kinetic reserve

The maximum power reserve created using inertial technique increases when the wind speed decreases as shown in Fig. 8. The maximum reserve obtained with this method is about 19 MW which is lower than the required reserve of 56MW. Due to the poor performance of the inertial technique, the aerodynamic method is used in this paper to create the power reserve. As noted in section *4.2*, this technique ensures the reserve regardless the wind speed and without any changes in the control strategy.

### 4.5. Power reference sharing between wind turbines and PMSG control to operate with power reserve

The method used for the power sharing between wind turbines is based on proportional distribution of the reserve as illustrated in (15). In this equation, $P_{tr\text{-}i\text{-}avail}$ presents the maximum power of the turbine located on line *i*, $P_{p\text{-}avail}$ is the maximum power available in the wind farm, $P_{r\text{-}ref}$ presents the required power profile for the grid which is imposed by the grid codes. The power reference distribution method is illustrated in Fig. 12.

$$P_{tr-i-ref} = \frac{P_{r-ref}}{P_{p-avail}} . P_{tr-i-avail} \tag{15}$$

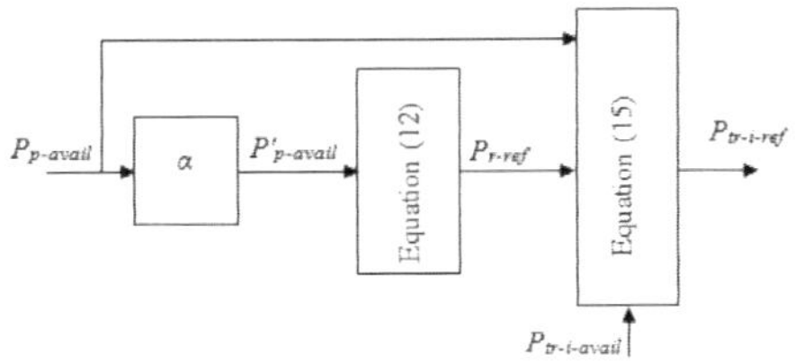


**Fig. 12.** Reserve distribution strategy in the wind farm.

The PMSG control strategy is based on the field-oriented control of the stator voltages [18]- [21] as illustrated in Fig. 13a. For the wind turbines operation with power's reserve, PMSG speeds and the wind turbines power are controlled through pitch angle estimation method as illustrated in Fig. 13a. These control loops are the same for all the wind turbines in the same wind farm.

The first role of the inverter presented in Fig. 13b is to maintain the DC-bus voltage at a constant value. The second role is to control the reactive power exchanged with the grid. More information about the control strategy and PI controller parameters calculation can be found in [18]-[21].

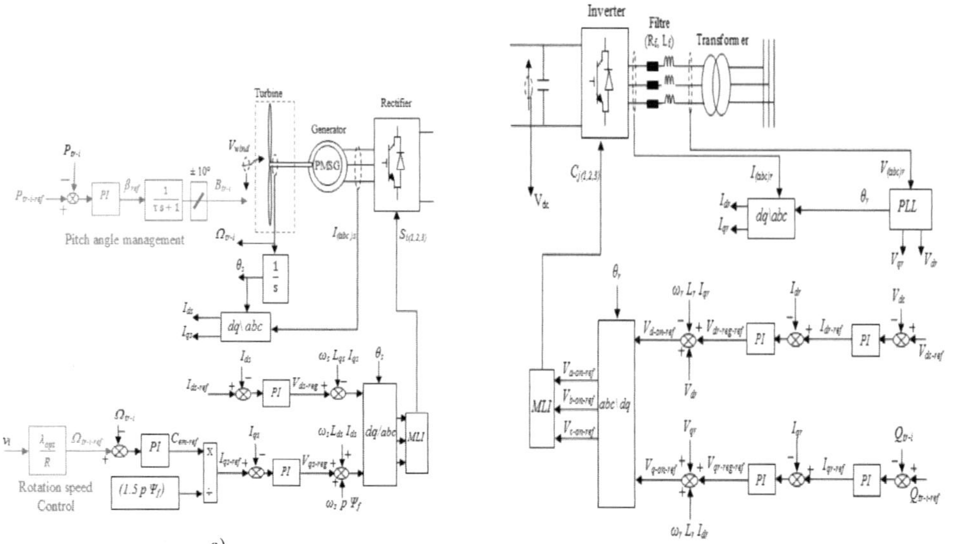


a) b)

**Fig.13.**a) Wind generator control strategy b) Energy management method between PMSG and the grid.

## 5. Simulation results

The system simulation is done using Matlab\SimPowerSystem tools with the parameters given in appendix section. The contributions of three wind turbines located in lines 1, 5 and 10 are respectively shown in Fig. 14, 15 and 16. The curves are obtained using wind generator control strategy illustrated in Fig. 13b for each Permanent Magnet Synchronous Generators (PMSG) in the wind farm. The figures 14$^{th}$ to 16$^{th}$ show that the strategy control proposed gives a good expectation as the power controlled are close to the reference power. The difference between the potential power of the three wind turbines located in lines 1, 5 and 10 compared to the power respectively produced are presented in Fig. 17, Fig. 18 and Fig. 19. We can notice that, the power's reserve extracted from the wind turbine located in line 1 is larger than those of line 5 and 10 and the same thing can be seen for the wind turbines on line 5 and 10. This difference is due to the method used for the power sharing between the wind turbines which takes into account the wind speed for each wind turbine.

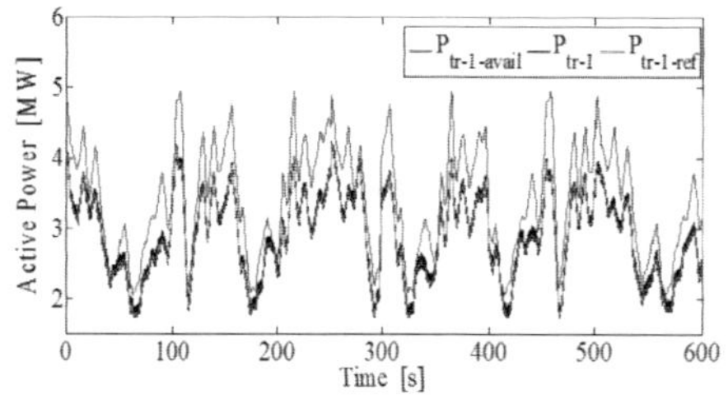


**Fig. 14.** Potential power of wind turbine located in line 1 Ptr-1-avail compared to its real production Ptr-1.

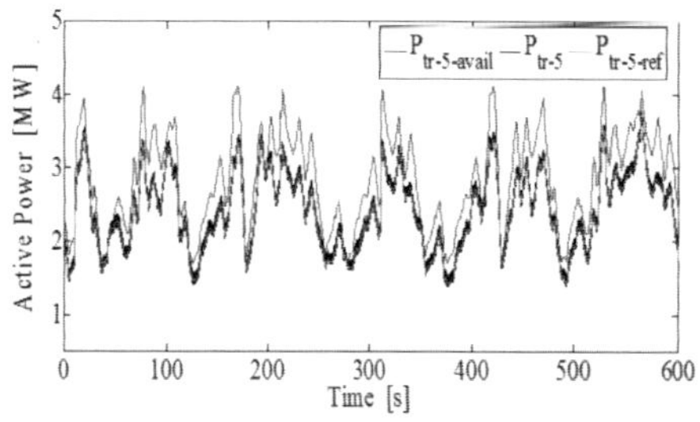


**Fig. 15.** Potential power of wind turbine located in line1 Ptr-5-avail compared to its real production Ptr

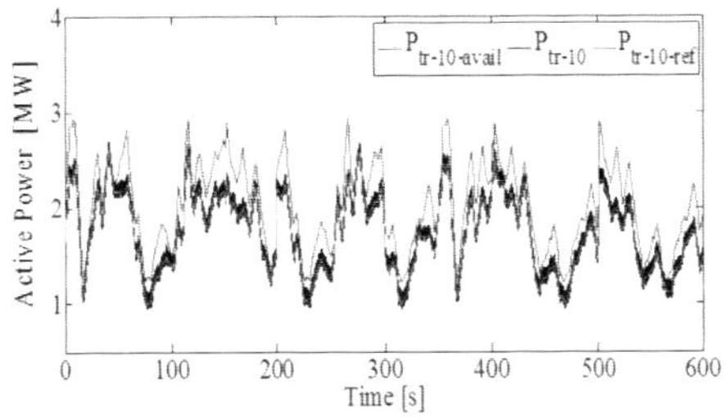


**Fig. 16.** Potential power of wind turbine located in line 1 Ptr-10-avail compared to its real production Ptr-10.

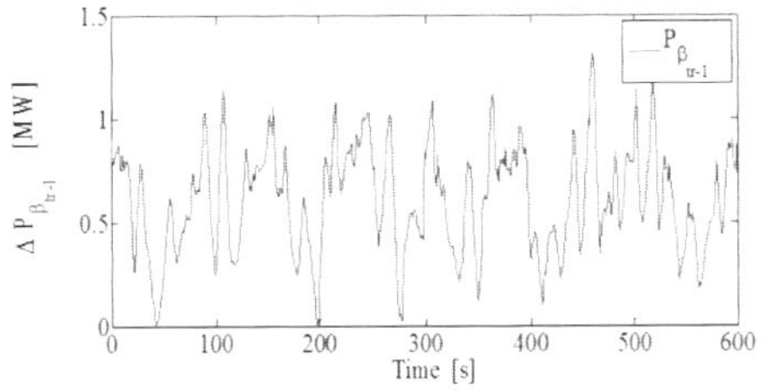


**Fig. 17.** Power's reserve extracted from wind turbine located in line 1.

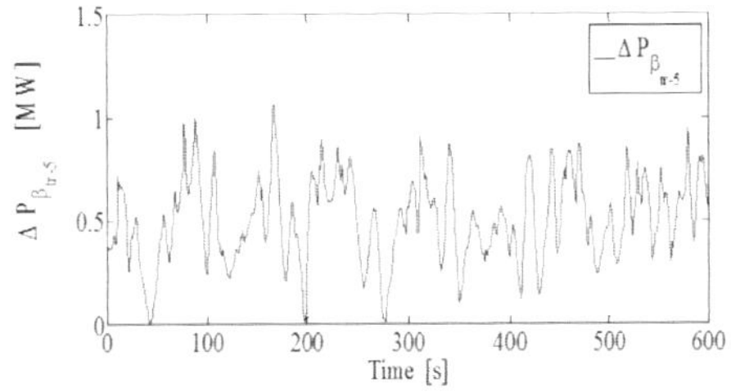


**Fig. 18.** Power's reserve extracted from wind turbine located in line 5.

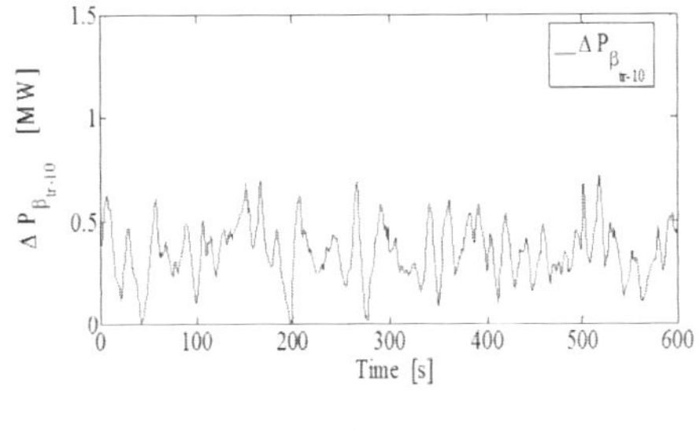


**Fig. 19.** Power's reserve extracted from wind turbine located in line 10.

The variations of the pitch angle for the wind turbines located in line 1, 5 and 10 are presented in Fig. 20. To analyze this last one, γtr-i is defined as the ratio between the output power and the available power for each wind turbine as expressed in (16). The γtr-i values for the three wind turbines are shown in Fig. 21. Its evolution is the same for the three wind turbines because the pitch angle variations are the same.

$$\gamma_{tr-i} = \frac{P_{tr-i-ref}}{P_{tr-i-avail}} \tag{16}$$

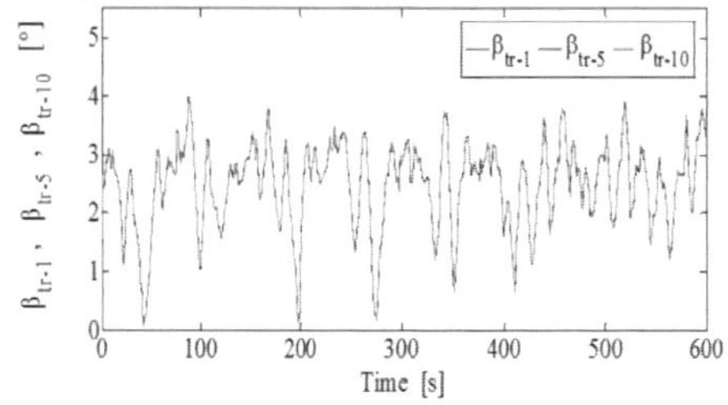


**Fig. 20.** Pitch angle variations for wind turbines located in line 1, 5 and 10.

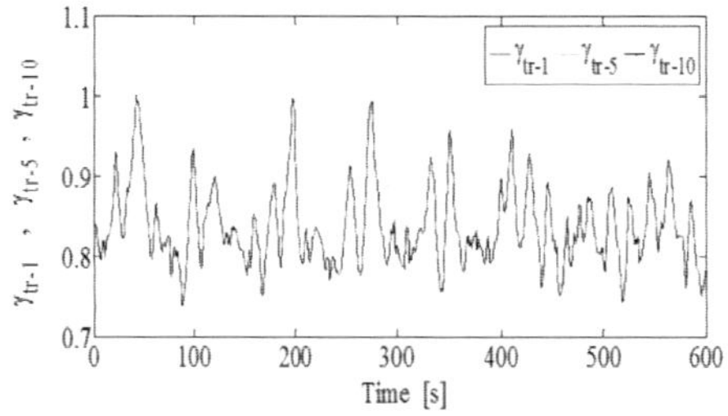


**Fig. 21.** Variation of ratio γ, γtr-1 for wind turbine 1, γtr-5 for wind turbine 5 and γtr-10 for wind turbine 10.

The power injected into the grid by the wind farm is presented in Fig. 22. This profile is adequate to the grid code requirements thank to the power's reserve based on aerodynamic approach performances as is illustrated in Fig. 23. The figure shows that the available power Pp-avail is more fluctuating compared to the power injected into the grid Pr. In other terms, the aerodynamic reserve enables the wind turbines to condition their productions so that, the wind farm production becomes acceptable to the grid code requirements.

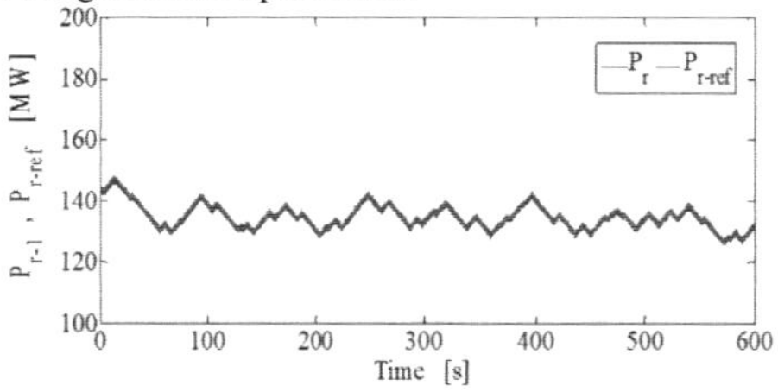


**Fig. 22.** Power injected into the grid from the wind farm.

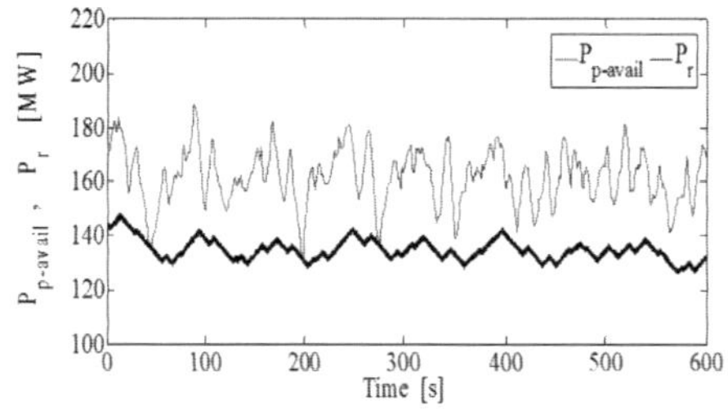


**Fig. 23.** Available power in the wind farm Pp-avail compared to the power injected into the grid Pr.

## 6. Conclusion

In this paper, a wind distribution model in the wind farm is proposed and evaluated by comparing the simulation results to the real data extracted from literature. Subsequently, an analysis of the quality of the energy produced by a wind farm is done with the goal to compare the rate of power fluctuations to what is currently admitted by E-on Germany grid code used in this study as a reference. The comparison is done by using the FHC factor.

The wind turbines operations with power's reserve enables to produce energy with the desired fluctuations rate. To create this reserve, an analysis of the wind turbine ability shows that there are two possible solutions. The first is based on inertial approach and the second is focused on aerodynamic method through the pitch angle estimation. The comparison of the two approaches shows that, the aerodynamic method is the best. The power reserve sharing between wind turbines is based on proportional distribution algorithms. The simulation results shows that, the strategy proposed is adequate. In other words, the wind farm is able to smooth its production by functioning with power's reserve. The most disadvantages of this strategy is that the wind farm is under exploitation and besides, we have the mechanical fatigue of the turbine due to recurrent vibrations of the blades.

## References

[1] T.R. Ayodele, A.A. Jimoh, J.L Munda, J.T Agee," *Challenges of grid integration of wind power on power System grid integrity*: A Review", International journal of renewable energy research (IJRER), **Vol.2**, **No.4**, (2012).

[2] J.Wilkes, I.Pineda, G. Corbetta, " *Wind energy scenarios for 2020*", European wind energy association (EWEA), technical report, (July 2014).

[3] RTE, Syndicat des énergies renouvelable, Enerdis, ADEef, *"Panorama de l'électricité solaire au 31 mars 2017"*, (March 2017).

[4] J. Kaboouris and F. D. Kanellos, " *Impacts of large-scale wind penetration on designing and operation of electric power systems"*, Sustainable Energy, IEEE Transactions on, **Vol. 1**,**No.2**, pp.107-114, (2010).

[5] Ministry of Ecology, Energy, Sustainable Development and the Sea, " *National Action Plan for renewable energies* (in French), **Article 4** of Directive 2009/28 / EC of the European Union, (2009).

[6] I.Pineda, S.Azau, J.Moccia and J.Wilkes, *"Wind in power 2013: European statistics"*, European wind energy association (EWEA), technical report, (February 2014).

[7] Ministry of Ecology, Sustainable Development and Energy, *"Wind power and photovoltaics: energy issues, industrial and societal (in French)* ", technical report, (September 2012).

[8] M. Tsili and S. Papathanassiou, " *A review of grid code technical requirements for wind farms"*, IET Renewable Power Generation, **Vol.3**, **No.3**, pp 308–332, (2009).

[9] A. H. Kassem, El-Saadany E.F, El-Tamaly H.H. and Wahab M.A.A., " *Ramp rate control and voltage regulation for grid directly connected wind turbines*", Power and Energy Society General Meeting - Conversion and Delivery of Electrical Energy in the 21st Century, 2008 IEEE, Pittsburgh, PA, USA, July 20-24, (2008).

[10] X.Han, Y.Qu, P.Wang and I. Yang, " *Four-dimensional wind speed model for adequacy assessment of power systems with wind farms*", Power Systems, IEEE Transactions On, **Vol.28**, **No.3**, pp. 2978-2985, (2013).

[11] J.Park, K.H. Law, " *Cooperative wind turbine control for maximizing wind farm power using sequential convex programming"*, Energy Conversion and Management,**Vol.101**, **No.1**, pp.295–316, (2015).

[12] H. K. Vladislav Akhmatov, " *An aggregate model of a grid-connected, large-scale, offshore wind farm for power stability investigations - importance of windmill mechanical system"*, International Journal of Electrical Power & Energy Systems, **Vol.24, No.9**, pp.709-717, 2002.

[13] C.E. Bănceanu, V.Iulian, " *Coordinated control of wind turbines"*, Master thesis, Department of Energy Technology - Pontoppidanstræde 101, Aalborg University, Denmark, June (2011).

[14] N. A. Kallioras, N.D. Lagaros, M.G. Karlaftis, P.Pachy, " *Optimum layout design of onshore wind farms considering stochastic loading*", Advances in Engineering Software, **Vol.88**, pp. 8–20, (2015).

[15] W. Li , G. Joos and C. Abbey, " *Wind power impact on system frequency deviation and an ESS based power filtering algorithm solution*", IEEE Power Systems Conference and Exposition, **PSCE '06**, 2006 IEEE PES, Atlanta GA, USA, October 29 -November 1, (2006).

[16] W.Li, " *An embedded energy storage system for attenuation of wind power fluctuations*", Ph.D. dissertation, Department Electrical Engineering, McGill University, Montreal, Canada, May (2010).

[17] E. Spahić and G. Balzer, " *The impact of the wind farm size on the power output fluctuation"*, European Wind Energy Conference (EWEC), Athens, Greece, February (2006).

[18] M.Singh, V.Gevorgian, E.Muljadi, E.Ela, " *Variable-speed wind power plant operating with reserve power capability"*, Energy Conversion Congress and Exposition (ECCE), 2013 IEEE, Denver, USA, September 15-19, (2013).

[19] M. Jannati, S.H.Hosseinian, B.Vahidi, Guo-JieLi,*"A survey on energy storage resources configurations in order to propose an optimum configuration for smoothing fluctuations of future large wind power plants*", Renewable and Sustainable Energy Reviews, **Vol.29**, pp. 158–172, (2014).

[20] Z. Zhou, F. Scuiller, J.F. Charpentier, M.E.H. Benbouzid and T. Tang," *Power smoothing control in a grid-connected marine current turbine system for compensating swell effect*", IEEE Transactions on Sustainable Energy, **Vol.4, No. 3**, pp. 816-826, (2013).

[21] D. Ikni, M.S. Camara, M.B. Camara , B. Dakyo, H. Gualous, " *Permanent Magnet Synchronous Generators for Large Offshore Wind Farm Connected to Grid - Comparative Study between DC and AC Configurations*", International journal of renewable energy research (IJRER), **Vol.4** ,**No.2**, (2014).

# Comparative study of geometrical configuration at the thermal performances of an agricultural greenhouse

*Lalmi* Djemoui[1,*], *Hocine* Bensaha[1] , *Abdelouahab* Benseddik[1] , *Ridha* Zarrit[1], *Mawloud* Guermoui[1], *Abdelaziz* Rabehi[1] and *Mohamed Tahar* Bouzaher[1]

[1] Unité de Recherche Appliquée en Energies Renouvelables, URAER, Centre de Développement des Energies Renouvelables, CDER, 47133, Ghardaïa, Algeria

**Abstract.** The aims objective of this work consists to study the storage system effects on the thermal performance of a tunnel agricultural greenhouse. The study focus on the use of the data climate analysis to predict the outside needs as comparison with another without storage system. The obtained results indicate that the outside needs are less than the no heated with 3 to 5°c during winter night. The thermal behavior of the greenhouse was study numerically and the results are corroborating with the literature. In addition, we conducted a comparative study designed to identify the optimal form of the greenhouse; two geometrical configuration are considered.

## 1 Introduction

The greenhouse is a production tool used to create artificially a microclimate to promote growth and development of culture. By its presence, it protects vegetation from blows due to changes in climate factors [1]. The main greenhouse environmental factors, which are different from the outside, are temperature, light and humidity. Each of these factors is conditioned in the greenhouse, by its level outside the enclosure, by the properties of the roofing material and by the characteristics specific to this greenhouse [2-3].

Tunnel-type plastic greenhouses are widely used around the world, especially in the Saharan countries because of their low investment cost [4].

These are efficient in winter and spring, where solar energy is useful and sufficient for greenhouse production. On the other hand, these greenhouses lose their effectiveness in summer where, the climate is very hot, which causes excessive overheating and strong hygrometries inside [5].

These extreme weather conditions affect the quality and quantity of the product and promote the development of certain diseases. On the physical plane, the greenhouse is a complex energy system in which all the different modes of thermal and mass exchange . If

* Corresponding author: eldjemoui@gmail.com

they are relatively simple and well known, their coupling causes difficulties in the modeling of the system [6-7].
In this system, natural convection is a particularly important mechanism for heat exchange between indoor air and all other solid surfaces (floor, walls, roof, culture, air conditioning and heating systems)[8].The aim objective of this study consists predict the storage system effect at the thermal behavior of an agricultural greenhouse in semi-arid climate . The study base on the use of the date climate to ameliorate the efficient of the production in this region with excess temperature. The outline of paper is: We start by an introduction; the potential solar is treated in the second parts. The third parts reserved to the description of the experiment. In fourth parts, we interpreted the results and discus and we finish by a conclusion and some perspectives.

## 2 Description of the experimental potential solar energy area study

In this study, experiments were carried out in two greenhouse at the Applied Research Unit on Renewable Energy at Ghardaïa from Algeria.Around of 77% of Algerian area presented arid and semi-arid regions. The characteristics of this region (Ghardaïa) are:

- Location 595Km south of the Mediterranean sea
- Latitude and 32°36 N
- Longitude3°80E
- Altitude of 469 m above the sea level
- Rate of sunny days per year: 77%
- Annual daily average of global solar irradiance about 7 kWh/m$^2$at horizontal surfaces.

Fig.1 show the variation of the average monthly global solar radiation in a period of 2016 [2-3].It can see that the global solar flux has the same trend of variation as the ambient temperature Fig.2 [1]. It is observed that the peak of average monthly radiation is registered in May, June, July and august (6000W/m$^2$) with a highest monthly ambient temperature surrounding 38°C. The least average monthly radiation is 2900W/m$^2$ with a monthly ambient temperature around 10° C in January and December. Figs.3and 4 gives the variation of the instannious monthly ambient temperature and relative humidity in four season of period of 2016. We can see that the ambient temperature passed the 40°C in august where the relative humidity passed 85% in January. An experimental study has carried out during the month of January 2016, which is the winter season.

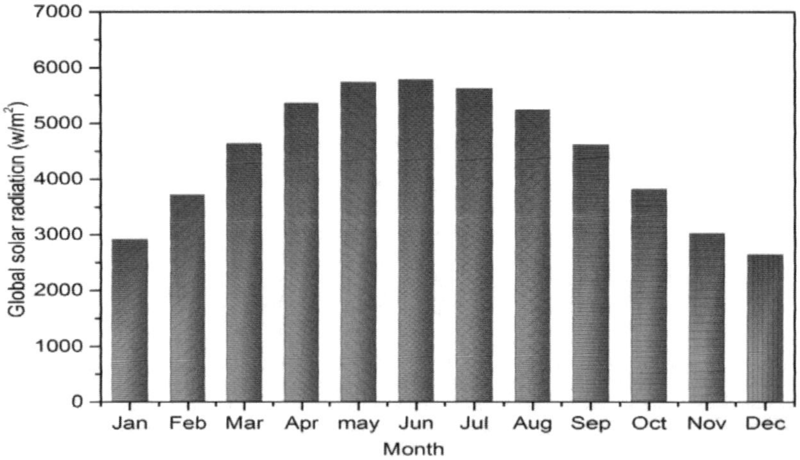

**Fig. 1.** Monthly variation of the global solar radiation for the year 2016 in Algeria.

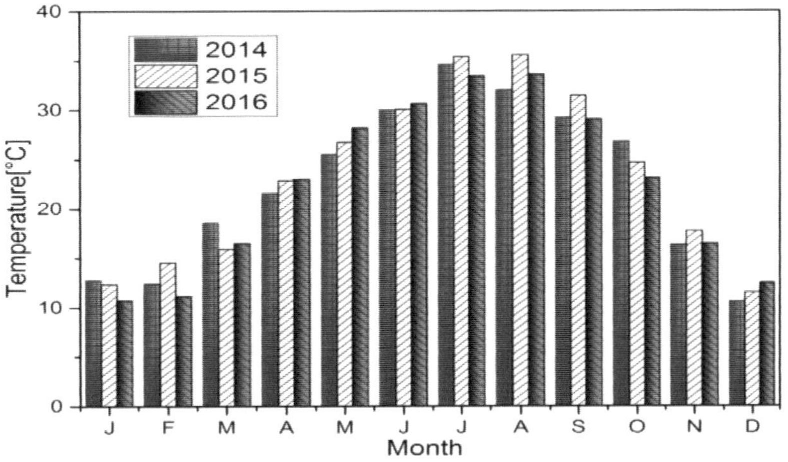


**Fig. 2.** Monthly variation of the mean ambient temperature in three years at Ghardaïa.

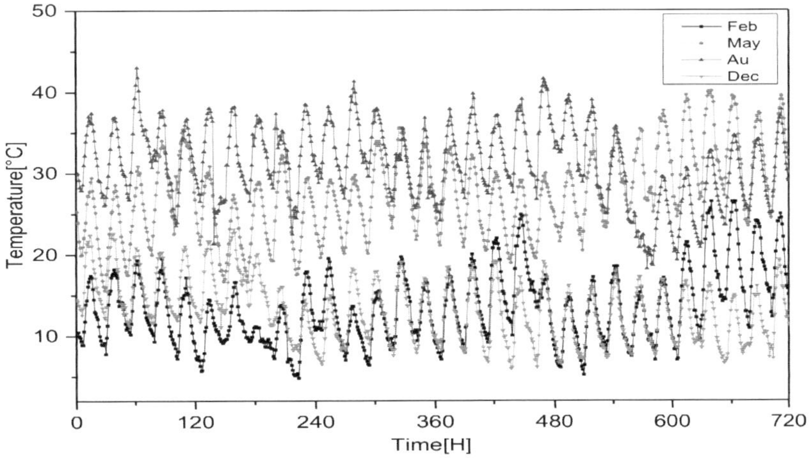


**Fig. 3.** Instannious temperature evolution in four month Feb, May, Aug and December2016.

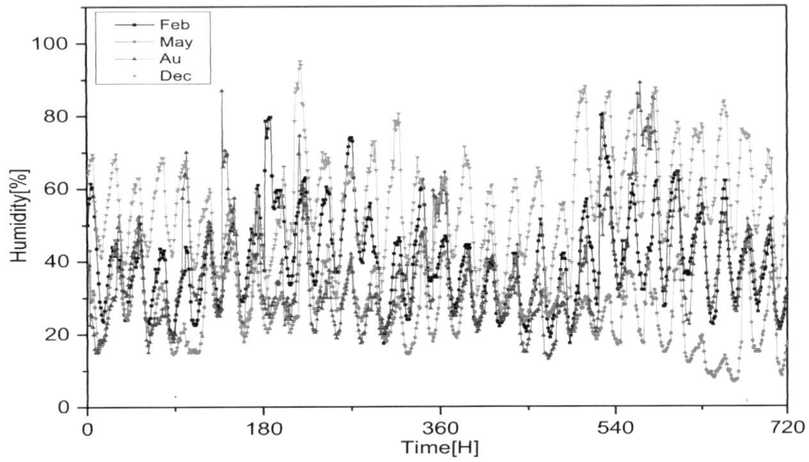

**Fig. 4.** Instannious relative humidity evolution in four month Feb, May, Aug and December 2016.

## 2.1 Experimental investigation

The two tunnel greenhouse test bench complete occupied and without storage system are shown in the Fig.5. It has a height of 3m, 25m of length and 8m of width, which leads to a volumetric of 528m$^3$.The greenhouse has a north-south direction, with optimal deviation angle of 32° to the West. It has doors (opening) at the side walls create for ventilation [11-12]. The polyethylene covers the experimental greenhouse with a low density, extensible and a 0.18 mm thickness. Polyethylene had a low and diffusion temperature respectively about -110°C and140°C.

**Fig. 5**. The two greenhouses test facility in URAER

## 2.2 Description of the storage system

Design steps of thermal storage system are presented in the Fig.6 and 7.It consists of the following parts:

- Four (4) PVC cylindrical pipes of 200 mm of diameter, implemented on the H position form.
- These pipe fields by stones
- Stones having 50 to 100 mm of approximate diameter.
- The storage system implemented about 0,70m away from the depth.

https://doi.org/10.1051/e3sconf/20186100003

**Fig. 6**. PVC pipe rumpled with stones

**Fig. 7**. PVC Emplacement

## 3 Experimentation and results

In the two greenhouses with and without storage system, courgettes plants are transplanted on the 27 th of December 2017 and the harvesting started on the 14 th of March 2018. The major environmental factors that affect the growth and the precocity of the production of greenhouse plants are temperature, light and humidity. The courgettes plants have two optimum temperatures, one during the day; which varied between 24 and 32 °C, and the other nocturnal temperature that is the most crucial temperature varied between 16 and 19 $^{0}$C. A courgettes crop (local variety) was planted in both greenhouses, planted at 20 plants in the row, and arranged in (8) eight rows with 30 cm between the rows and 20 cm between the plants. During the growing period, the January has the least average monthly radiation is 2600W/m$^2$ with a monthly ambient temperature around 10°C (Fig.2 and 3).

The heating requirements of both greenhouses with and without storage system at night, are shown in Fig.8 with a set temperature point of 17 ° C (required temperature for plants at night our case courgettes).

The curves show that the needs of the experimental greenhouse occupied with the storage system are lower than the without storage system with difference between 3 to 5°C per night.

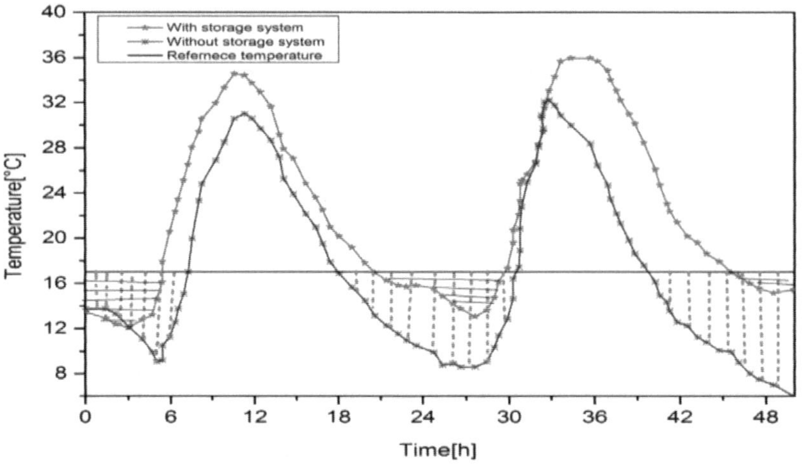


**Fig. 8**. Temperature evolution at night of both greenhouses (heating requirement).

## 4 Numerical study

The aim objective of this investigation is to study the air behavior evolution inside the heated greenhouse without vegetal cover by comparison with an author chapel greenhouse. The Ansys software 16.0 code used. Flow considered laminar in two-dimensional along x and y, where the Physical properties of air supposed constant, except its density (ρ) whose variation was given by the Boussinesq approximation [13-14-15-16]:

$$\rho = \rho_0(1-\beta(T-T_0)) \tag{1}$$

### 4.1 Geometrical configuration

Geometrical configuration with all boundaries condition considered in this study showed in the Fig.9.

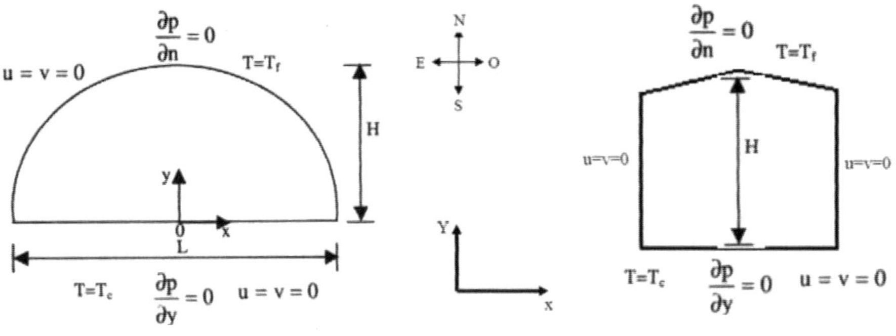


**Fig. 9**. Problem Position and boundary condition for tunnel greenhouse

### 4.2 Simulation results

Streamlines and isotherms for different Rayleigh numbers (Ra) are shown in Fig.10 and 11. Fluid flow intensifies and natural convection increases and predominates on conduction.

The heated air particles at ground level rise along the wall then, cooled particles in contact with roof goes approximately to the author walls.
The influence of Rayleigh number on the streamline (top the fig.10 and bottom the fig.11) and the isotherms (bottom of the Fig.10and top of the Fig.11) is illustrated. With different Rayleigh numbers the flow characterized by two air circulation loops in both cases. For small Rayleigh numbers (between Ra = 104 to 105), the isotherms are parallel. This representation indicate that the heat transfer dominated by conduction.

When the Rayleigh number increases, isotherms becomes more and more wavy and transfer heat exchange was more pronounced in the two cases, tunnel and chappel greenhouse. As this obtained result, fluid flow intensifies and natural convection increases and dominates conduction. The heated air particles at ground level rise along the wall. Then, the cooled particles in contact with the roof flow approximately a median plane.

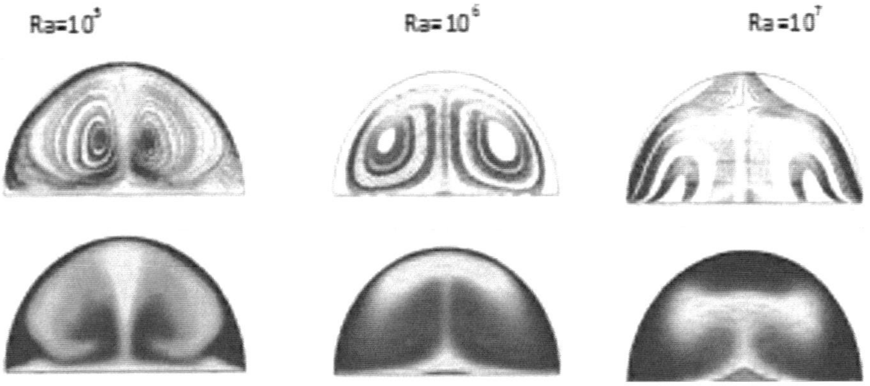

**Fig. 10.** Isolignes and isothermal evolution in tunnel greenhouse

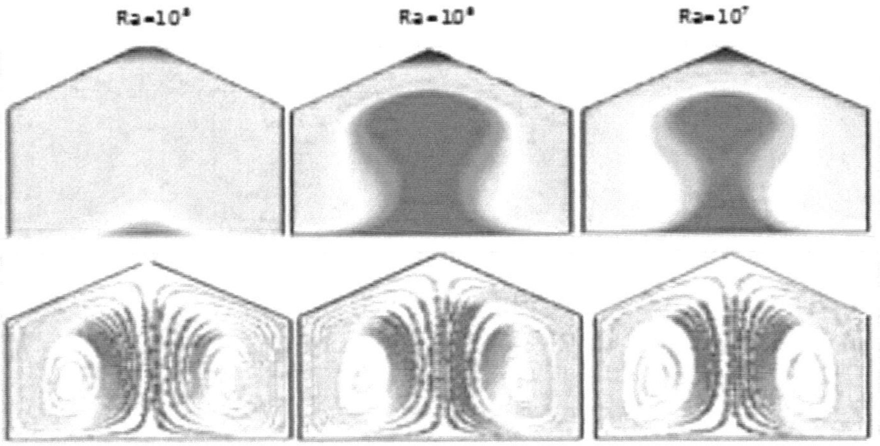

**Fig. 11.** Isolignes and isothermal evolution in Chappel greenhouse

## 5 Rayleigh number Effect's at the heat transfer exchange

The exchange heat transfer evolution in the both proposed geometrical configuration of greenhouse for different Rayleigh numbers represented by the local Nusselt number in Fig.12 and (13).Therefore, the logic was respected as long as there is a concentration of isotherms at the corners (ground), which explains a large number of Nusselt. It was found that for a small Rayleigh number varying between $10^4$ to $10^5$, the Nusselt number of is small and it is the conduction that dominates. With increase of Rayleigh number, the exchange rate increases and the local Nusselt number becomes important in the tunnel compared with the chappel greenhouse. The rate transfer represented according to the mean Nusselt number Fig.14. It can be see that the tunnel is higher losses compared to the chappel. Finally, from these results; we can observe the good according between the literature and the present work for different Rayleigh numbers.

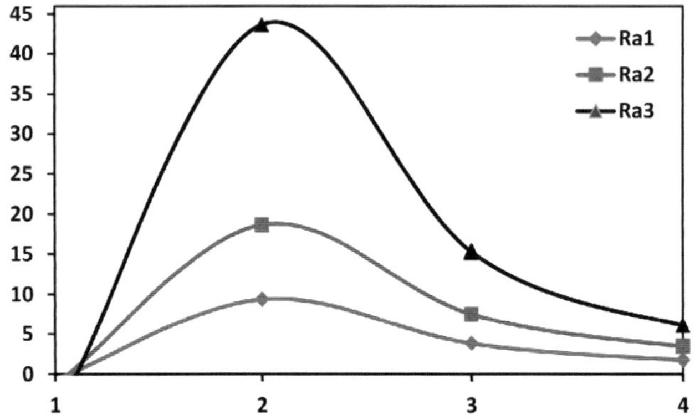


**Fig. 12**. Local Nusselt Number evolution in tunnel greenhouse at three values of Rayleigh number Ra=$10^5$, $10^6$and $10^7$

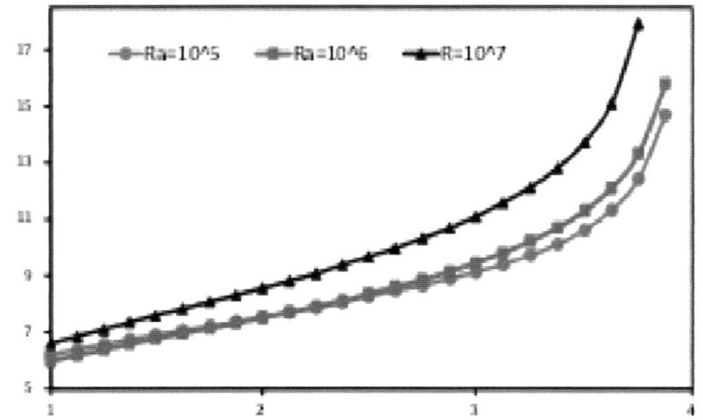


**Fig.13.** local Nusselt Number evolution in Chapel greenhouse at three values of Rayleigh number Ra=$10^5$, $10^6$and $10^7$

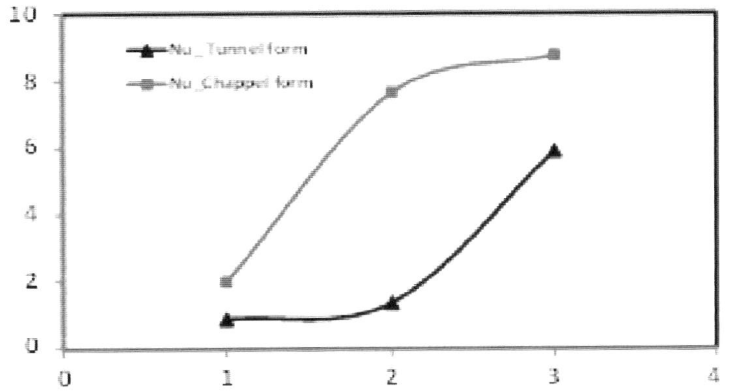


**Fig.14.** Mean Nusselt number evolution in tunnel and Chappell greenhouse

## Conclusion

Our objective was to study the storage system effects and the air behavior inside the greenhouse.The thermal performences of the heated grenhouse are ameliorated with outsit needs between 3 to 5 °C.Numerical similation was conducted to predict the air behvior inside the grenhouse by the spatiotemporal distributions of isolignes and isothermal.We have also shown that for the conditions imposed on the and for low differences temperature between floor and roof, the air circulation is characterized by two recirculation cells rotating in the opposite direction. Therefore, this study should make it possible to improve the thermal design of greenhouse as well as the positioning of heating systems with thermal storageand geometrical configuration.

## References

1. D. Lalmi, S. Bezari, H. Bensaha, M. Guermoui, A. Rabehi, B. Abdelouahab, R. Hadef, Analysis of thermal performance of an agricultural greenhouse heated by a storage system, Modelling, Measurement and Control B, (2018)
2. Abdelaziz Rabehi, Mawloud Guermoui, Lalmi Djemoui, Hybrid Models for Global Solar Radiation Prediction: case of study, International Journal of Ambient Energy, (2018).
3. Reski Khelifi ,Guermoui Mawloud, Rabehi Abdelaziz, Lalmi Djemoui , Multi-Step Ahead Forecasting of Daily Solar Radiation Components in Saharan Climate, International Journal of Ambient Energy, (2018)
4. Naijun Zhou „ Yaxiong Yu , Jinping Yi , Rui Liu ,A study on thermal calculation method for a plastic greenhouse with solar energy storage and heating ,Solar Energy **142** (2017) 39-48
5. Nessim Arfaoui, Salwa Bouadila, Amenallah Guizani ,A highly efficient solution of off-sunshine solar air heating using two packed beds of latent storage energy , Solar Energy **155** (2017) 1243–1253
6. M. Lazaar , S. Kool P, M. Hazam P, A. Farhat a, A. Belghith , Use of solar energy for the agricultural greenhouses Use of solar energy for the agricultural greenhouses autonomous Conditioning, Desalination **168** (2004) 169-175

7. T. Boulard , S. Wang, Greenhouse crop transpiration simulation from external climate conditions, Agricultural and Forest Meteorology **100** (2000) 25–34
8. A.Bouhdjar, M. Belhamel, F.E. Belkhiri and A. Boulbina, Performance of sensible heat storage in a rock bed used in a tunnel greenhouse. WREC 1996
9. F. Berroug, E.K. Lakhal, M. El Omari, M. Faraji and H. El Qarnia, Numerical Study of Greenhouse Nocturnal Heat Losses, Journal of Thermal Science Vol.**20**, No.4 (2011) 377
10. Dewanto Harjunowibowo, Erdem Cuce, Siddig A. Omer, Saffa B. Riffat, Recent Passive Technologies of Greenhouse Systems-A Review, 15th International Conference on Sustainable Energy Technologies – SET 2016, 19 th -22nd of July 2016, National University of Singapore, Singapore
11. Erdem Cuce, Dewanto Harjunowibowo, Pinar Mert Cuce ,Renewable and sustainable energy saving strategies for greenhouse systems: A comprehensive review, Renewable and Sustainable Energy Reviews 64 (2016)34-59.
12. George C. Bakos, Dimitrios Fidanidis, Nikolaos F. Tsagas, Greenhouse heating using geothermal energy, Geothermics 28 (1999) 759-765
13. Bezari S, Lalmi D, Babbou A. (2016).Heating of greenhouse with storage system: study and investigation. 4th international conference on Renewable Energies (CIER_2016) Proceedings of Engineering and Technology 14: 23-28.
14. Babbou A, Ziari YK, Lalmi D, Bezari S, Zarit R, Bensaha H.. Effect of operational parameters on the performance of heat storage in a rock bed used in a tunnel greenhouse. The 4th International Seminar on New and Renewable Energies, Ghardaïa – Algeria (2016).
15. Lalmi D, Hadef R, Babbou A, Bensaha H, Bezari S, Zarit R. Numerical study of air behavior in a greenhouse Equipped with a thermal storage system. The 4th International Seminar on New and Renewable Energies, Ghardaïa – Algeria (2016).
16. D.Lalmi & R.Hadef, Numerical simulation of co and counter swirls on the isothermal flow and mixture field in a combustion chamber, Advances and Applications in Fluid Mechanics, AAFM, Vol. **18**, issue 2:199-212

***Nomenclature***

*Ra* *Rayleigh number*
*Ra* *Reynolds number*
*Nux* *Local Nusselt Number*
*Num* *Mean Nusselt Number*
*Pr* *Prandtl number*
*Re* *Reynolds number*
*Pvc* *Plastic pipe*
$I_t$ *total solar radiation (W=m2)*
*A* *surface area* ($m^2$)
*m* *mass(kg)*
*Q* *heat rate (W)*
*t* *time (s)*
*hi* *Convective heat transfer coefficient between liquid and*
*P* *Pressure (pas)*
*Cp* *specific heat of air at constant pressure (J/kgK)*
*T* *Temperature*
*U* *velocity (m=s)*
*V* *volume* ($m^3$)
*Y* *depth of soil (m)*
*M* *mass flow rate* ($kg/s^{-1}$)
$D_0$ *Diameter (m)*
*D* *diffusion coefficient*

***Exposants***

*Cd* *conductive heat*
*Cv* *convective heat*
*L* *latent heat*
*R* *radiation heat*

***Subscript***

*Max* *maximum*
*Min* *minimum*
*H* *heat gain*
*A* *absorbed*
*a* *ambient*
*co* *cover of greenhouse*
*b* *bottom*
*e* *edge*
*to* *top*
*pol* *polyethylene*
*av* *average*
*inl* *inlet greenhouse*
*in* *inside greenhouse*
*out* *outside greenhouse*
*so* *soil*
*sts* *storeg system*

***Greek Symbols***

$\varepsilon$ *emissivity*
$\Delta$ *difference in time*
$\lambda$ *Thermal conductivity (W/mK)*
$\alpha$ *thermal diffusivity*
$\mu$ *dynamic viscosity* ($Ns/m^2$)

# Modeling of Hybrid Solar Palel For Producing Hydrogen

*Chawki* AMEUR MENAD[1,*], and *Rabah* GOMRI[1]

[1] University Frères Mentouri Constantine, Faculty of Sciences and Technology, Department of Génie Climatique, 25000 Constatine, Algeria

**Abstract.** Applied renewable energy for improving the quality of life has taken the major part of the modern research. The purpose of doing this research is developing new technology which is modeling a new hybrid solar panels generating by different sources of renewable energy, and mini mechanical systems. As a result generating hybrid solar panel by geothermal energy and solar energy, including the mini-turbine increases the water's temperature from 95°C to 622°C for producing the hydrogen through the electrochemical thermodynamic by using HTES Technology.

## 1 INTRODUCTION

A lot of research projects have been developed about exploiting the energy in different ways. However; the renewable sources became the most important field of research due to the availability and impact on the environment. Few examples about using solar systems for hydrogen were proposed. Acakpovi et al [1] have studied modeling solar power generation by photovoltaic cell, where the modeling was under Simulink, the obtained results confirmed some novelty comparing to the previous research. Another research have published about design and studying new technique called Support Vector Machine (SVM) based on experimental experience. Where Kazem et al [2], obtained results proofed that the studied model achieved less MSE in comparing to the previous research about management energy generation. Moreover, Salmi et al [3] have developed a new model photovoltaic cell by using Matlab/SIMULINK. The model is based on mathematical equations and described through an equivalent circuit including a photocurrent source. In addition, an experimental test bench was built and the obtained results exhibited a good agreement with the simulation ones. Regarding studying novel modeling technique of PV module with a fuzzy logic based MPPT algorithm and boost converter in Simulink environment which was developed by Mahamudul et al [4]. The main result has given an important contribution which is simplification of PV modeling technique and implementation of fuzzy based MPPT system to track maximum power efficiently. Other research about developing methodology for production of hydrogen by electrolysis system was proposed by Ayati et al [5] where the electricity is delivered by photovoltaic panel. Wilson et al [6] have mentioned in their research the objective of generating Hydrogen by solar photovoltaic (PV) array and

* Corresponding author: c.ameurmenad@gmail.com

then collected for data analysis to demonstrate efficiency of the hydrogen production in all the steps of the experiment. Their positive results proofed that the studied case model allows demonstrating at least one hour operation per hydrogen charging at room temperature. Agrafiotis et al [7] have been proposed the exploitation of solar energy for the dissociation of water and production of hydrogen via an integrated thermo-chemical reactor/receiver system. They obtained results which have explained that the operation of the complete cycle by a single solar energy converter. Some researchers were focusing just in production of hydrogen by electrolysis powered from wind and solar sources. As a result they got the reasonable price for producing hydrogen from wind and solar energy [8]. Other choice for exploiting solar power for creating new opportunities for hydrogen production, where the final cost estimation for hydrogen and solar electricity generation will show the potential of both technologies for a sustainable energy industry, this research has been developed by Volker Quaschning and Franz Trieb [9]. Finally, Crook et al [10] have developed a new simulation method for creation a hybrid panels by combination between solar thermal energy and photovoltaic energy and phase change material system; they considered the application of wind energy for production of electricity.

## 2 DESCRIPTION OF THE HYBRID SOLAR PANEL

The proposed hybrid solar panel is based on combining between three different sources of energy, which are Silicon solar cells, mini-turbine and geothermal source of energy. The panel absorbs the solar radiation and transforms it to electricity through solar cells. Moreover the geothermal source passes through channel which is rounding by electrical resistance for increasing the vapor water temperature for hydrogen production by using the electrochemical thermodynamic by using HTES Technology. And the last part, which is the mini-turbine for increasing the amount of electricity producing by hybrid solar panel. The studied case consists modeling two different parts; the first one is thermal part which combines between geothermal energy and mini-mechanical system (mini-turbine). The second part consists modeling photovoltaic cells. Coupling between the two parts gives us the total power produced by the hybrid solar panel. The aim of studying the proposed case with mini turbine is to gain more amount of electricity and increasing of the temperature of vapor water for production of electricity and hydrogen.

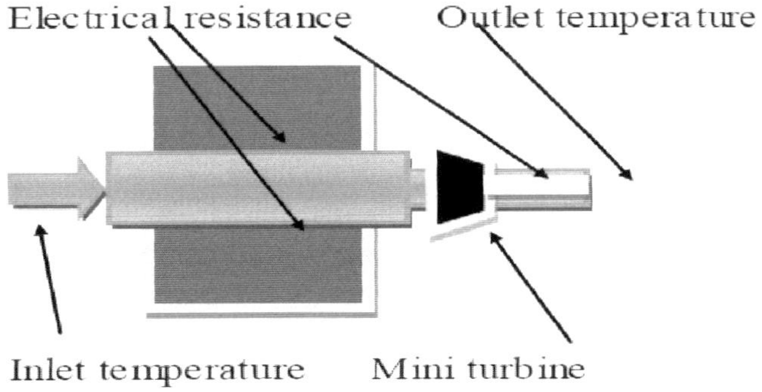


Fig. 1. Coupling between the photovoltaic energy and geothermal energy for production of electricity with mini turbine and increasing of vapor water.

# 3 MODELING OF THE HYDRIDE SOLAR PANEL

## 3.1 Electrical part

In this part; understanding the Ideal Solar Cell (simplified), is the aim for getting the maximum Power producing. The voltage current characteristics equation of PV Ideal solar cell simplified is given by:

$$\mathrm{I} = Iph - I_0.(exp\left(\frac{q.v}{n.k.T}\right)) \quad (1)$$

Where the photon current in solar Ideal cell is:

$$\mathrm{Iph} = \left(IS + Ki(T - Tn)\right)\frac{G}{Gn} \quad (2)$$

Where *Ki*=0.0017 A/◦C is the cell's short circuit current temperature coefficient and *G* is the solar radiation (W/m2).
The cell's saturation current varies with temperature which is:

$$\mathrm{Is(t)} = Is.(\frac{t}{tn})^3.exp(\frac{Eg}{Vt}\left(\frac{t}{tn} - 1\right)) \quad (3)$$

Where *Is* is the diode reverse saturation current, *tn* is the nominal temperature, *Eg* is the band gap energy of the semiconductor and *Vt* is the thermal voltage.
Vt is called thermal voltage:

$$Vt = \frac{k.T}{q} = \frac{T}{11600} \quad (4)$$

The current output of PV module is:

$$I = Np.Ip - NpI0(\exp\left(\frac{q.v}{n.k.T}\right)) \quad (5)$$

## 3.2 The thermal part

The electrical power produced by the mini turbine is shown in the following equation:

$$\mathrm{Pu} = 0{,}97.9{,}81.\mathrm{nt}\,.\,\mathrm{ng}\,.\,\mathrm{Qt}\,.\mathrm{Hn} \quad (6)$$

Where; Pu, Qt, Hn, nt, ng represent respectively the amount of energy produced (W), the flow of the mini turbine (m3/s), net drop (m), the efficiency of the mini turbine, efficiency of the generator (generator of the mini turbine). The following equation show the net drop :

$$\mathrm{Hn} = \mathrm{Hb} - \mathrm{HL} - \mathrm{V}^2/\,2.\mathrm{g} \quad (7)$$

Where, Hb, HL, v, g, represent respectively Gross drop, higher, celerity, cceleration (HL represent about 10% To 15% of Hb). The outlet temperature is depicts in the following equation:

$$\mathrm{Ts} = \mathrm{Te} + (\mathrm{Pu}\,/\mathrm{m.\ Cp}) \quad (8)$$

Where Ts, Te, m, Cp, Pu: are respectively outlet temperature, inlet temperature, specific heat, electrical power of vapor water between the inlet and outlet point of each channel connected on the solar panel.

### 3.3 Programming the hybrid solar panel by using FORTRAN 95

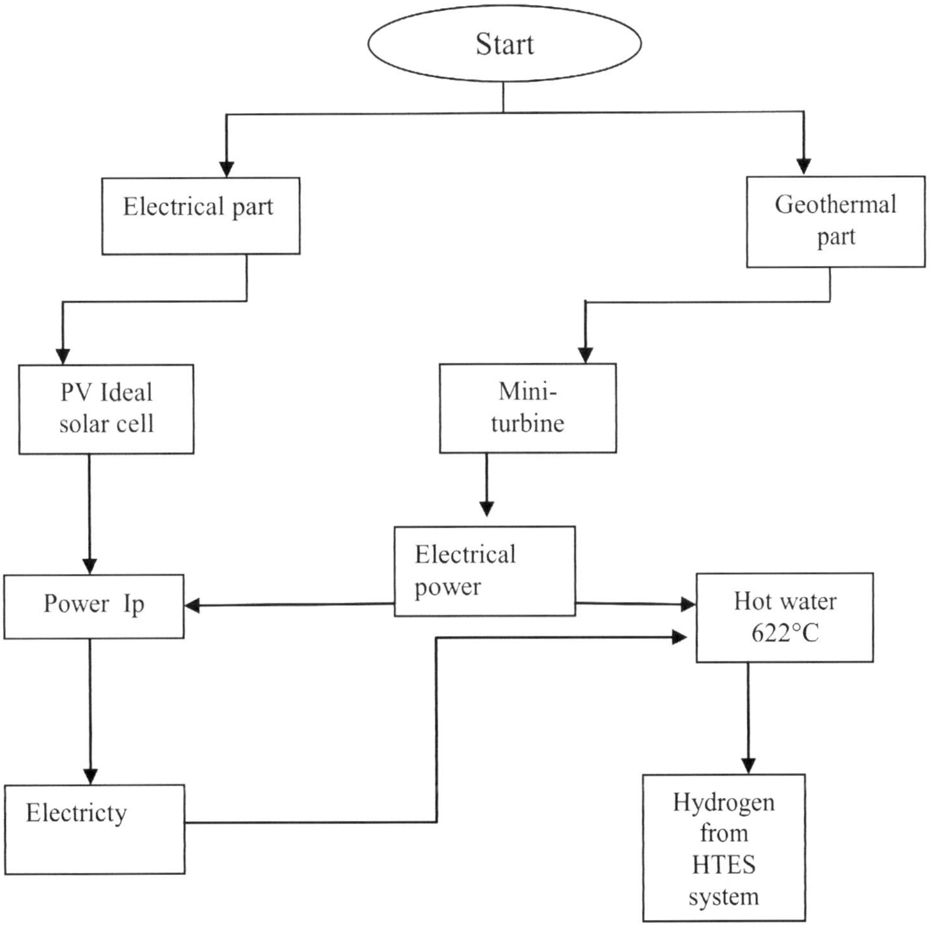


Fig.2. programming FOTRAN 95 for Hydride solar panel for hydrogen production

## 4 Hydrogen production

### 4.1 Description of hydrogen production from HTES system

Production of hydrogen from steam water at 622°C in the studied case is possible by using the HTES technology where the temperature varied between 600°C and 1000°C. However there is another possibility to increase the temperature until 1000 from the same hybrid solar panel, and start producing hydrogen from different such as:

$H_2SO_4$ (130°Cto 1000°C).

$$SO_2+I_2+2H_2O \rightarrow H_2SO_4+2HI \qquad (100°C\text{-}130°C) \qquad (9)$$

$H_2SO_4 \rightarrow SO_2+H_2O+1/2\ O_2$ (800°C-1000°C) (10)

$2HI \rightarrow H_2+I_2$ (300°C-400°C) (11)

$CH_4$ (700°C-1000°C) (2Bar-25Bar)

$CH_4+H_2O \leftrightarrows CO+3H_2$ (12)

$CO+H_2O \leftrightarrows CO2+H_2$ (13)

## 5 RESULTS AND DISCUSSIONS

The aim of the proposed model is just develop a hybrid solar panel for one goal which is producing hydrogen by using HTES technology, so the water vapor water should be varied between 600°Cto 1000°C, the following table resume the purpose of the research.

**Table1.** characteristics of hydride solar panel for hydrogen production.

| | Power (W) | Outlet Temperature (°C) | Number of panels obtained | Mini-turbine (W) |
|---|---|---|---|---|
| Hybrid solar panel | 1784 | 622 | 4 | 3000 |

The obtained results resumes the studied model which has given an electrical power of 1784W in 4 solar panel were connected in serial, however the most important factor is the high temperature of the vapor water for hydrogen production. Moreover; the power obtained by mini-turbine has increased the total amount of electricity, for focusing in one purpose which is hydrogen production.

In terms the variation of outlet temperature through the four hybrid solar panels; the following table (Table. 2) explains that the power obtained increases the temperature of the vapor water by transforming electrical energy to thermal energy through an electrical resistance around the channel. Increasing the power from 500W to 1784W has increased the temperature from 310°C to 622°C. As a result the proposal model is stable for producing electricity and instable for increasing the vapor water temperature. However; it achieves the exact temperature for producing hydrogen by using HTES system

**Table 2.** Variation of outlet temperature on function of Power produced by hybrid panel.

| Pu(W) | Ts (°C) |
|---|---|
| 500 | 310 |
| 1000 | 425 |
| 1500 | 545 |
| 1784 | 622 |

Finally; understanding the variation of speed of the vapor water is necessary for calculating the power produced by the mini-turbine. The Table 3 mentions that increasing the speed of vapor water improves the quality of the hybrid solar panel by increasing the power producing. In addition; increasing the speed from 5.10m/s to 16m/s has increased the produced power from 500W to 1784W so a difference of 1400W.

**Table 3.** variation of vapor speed on function of Power produced by hybrid panel..

| Pu(W) | V (m/s) |
|---|---|
| 500 | 5.10 |
| 1000 | 7.10 |
| 1500 | 10.10 |
| 1784 | 16 |

## Results of Hydrogen production

The obtained results show that the proposed hybrid solar panel is able to produce a considerable amount H2, However; CH4 cycle is the best cycle in high HTES systems.

**Table 4.** Amount of hydrogen produced from H2SO4 and CH4.

| Cycles for Hydrogen Production. | Hydrogen productivity. $MJ/KgH_2$ | Amount of Hydrogen production. |
|---|---|---|
| H2SO4.<br>$SO_2+I_2+2H_2O \rightarrow H_2SO_4+2HI$<br>$H_2SO_4 \rightarrow SO_2+H_2O+1/2\ O_2$<br>$2H \rightarrow H_2+I_2$ | 295,500MJ/KgH2 | 0,517KgH2/day |
| CH4<br>CH4+H2O $\leftrightarrows$ CO+3H2<br>CO+H2O $\leftrightarrows$ CO2+H2 | 165MJ/KgH2 | 0,9394KgH2/day |

## References

1. A. Acakpovi, E. Ben Hagan; Novel Photovoltaic Module Modeling using Matlab/Simulink, International Journal of Computer Applications (0975 – 8887) Volume 83 – No.16, (2013)
2. H.A. Kazem, J. H. Yousif, M.T. Chaichan, Modelling of Daily Solar Energy System Prediction using Support Vector Machine for Oman , International Journal of Applied Engineering Research ISSN 0973-4562 Vol. 11, Number 20, 10166-10172, (2016).
3. T. Salmi, M. Bouzguenda, A. Gastli, A. Masmoudi, MATLAB/Simulink Based Modelling of Solar Photovoltaic Cell, International Journal of Renewable Energy Research, Vol.2, No.2, (2012).
4. H. Mahamudul, M. Saad, M. I. Henk, Photovoltaic System Modeling with Fuzzy Logic Based Maximum Power Point Tracking Algorithm, International Journal of Photoenergy http://dx.doi.org/10.1155/2013/762946, (2013)
5. F. Ayati, A. M'Raoui, M. Belhamel, A. Rebai ; Modélisation d'un Système de Production d'Hydrogène Solaire par Electrolyse, Rev. Energ. Ren. Vol. 7, 135-150, (2004).

6. J.F. Wilson , S.S. Srinivasan, B. M. Moore, L. Henderson, S. Ely, P.C. Sharma; Hydrogen production using solar energy , Journal of Undergraduate Research in Physics, (2013).
7. A. Agrafiotis, M. Roeb, A.G. Konstandopoulos , L. Nalbandian, V.T. Zaspalis, C. Sattler, P. Stobbe, A.M. Steele, Solar water splitting for hydrogen production with monolithic reactors. Solar Energy, Vol. 79, 409–421, (2005).
8. A. Rahil, R. Gammon, Dispatchable Hydrogen Production at the Forecourt for Electricity Demand Shaping Sustainability, Vol.9, 1785, (2017)
9. Quaschning, F. Trieb, Solar Thermal Power Plants for Hydrogen Production, Hypothesis IV Symposium Stralsund, Germany 9-14 September, 198-202, (2001).
10. C.S. Malvi, D.W. Dixon-Hardy, R. Crook, Energy balance model of combined photovoltaic solar-thermal system incorporating phase change material, Solar Energy, Vol. 85, Issue 7, 1440–1446, (2011).

# Experimental Study of Numerical Relay for Over-current Protection in Solar Panel for Securing the Hydrogen production.

*Chawki* Ameur menad[1,*], *M.* Bouchahdane[2], and *Rabah* Gomri[1]

[1]University Frères Mentouri Constantine 1, Faculty of Sciences and Technology, Department of Génie Climatique, 25000 Constatine, Algeria

[2]Department of Power and control Institute of Electrical and Electronic Engineering, IGEE Boumerdes, Algeria

**Abstract.**. Every electrical system in solar panel can fail during electrical faults. In this incidence, high fault current can occur. Such current must be interrupted by a protective system. The research was supported by experimental tests. In work conditions close to real, the numerical relay REF542plus was tested for both instantaneous and extremely inverse definite minimum time IDMT over-current protection functions with the help of CMC 365 injection and test equipment associated to Test Universe software. Protecting hybrid solar panels generating by different renewable energy sources for hydrogen production from over-current is very important for improving the energy efficiency in one hand, and securing the function in critical condition from damage of the solar cells in second hand. The contribution of this research is controlling the over-current in the solar panel for securing the continuation of the hydrogen production from renewable energy sources in short time. The obtained results allowed the observation of the relay's behavior when subjected to certain faults; where the solar panel keeps producing the hydrogen.

## 1 INTRODUCTION

Due to the impact of the safety in solar systems on the energy efficiency in critical situation, an experimental proposal research project was studied for understanding the most important problem which is the over-current in solar systems in Algerian Climate. The solar has been installed for exploiting electricity produced for hydrogen production from H2SO4 cycle under Algerian climate. Many research projects have been developed about using renewable energy for hydrogen production. Wilson et al [1] have mentioned on their research the objective of generating Hydrogen by solar photovoltaic (PV) array and then collected for data analysis to demonstrate efficiency of the hydrogen production in all the steps of the experiment. Their positive results proofed that the studied case model allows demonstrating at least one hour operation per hydrogen charging at room temperature. Some research papers were focusing just in production of hydrogen by electrolysis powered

* Corresponding author: c.ameurmenad@gmail.com

from wind and solar sources. As a result they got the reasonable price for producing hydrogen from wind and solar energy [2]. Other choice for exploiting solar power for creating new opportunities for hydrogen production, where the final cost estimation for hydrogen and solar electricity generation will show the potential of both technologies for a sustainable energy industry, this research has been developed by Quaschning and Trieb [3]. Amevi Acakpovi, Essel Ben Hagan have studied modeling solar power generation by photovoltaic cell, where the modeling was under Simulink, the obtained results confirmed some novelty comparing to the previous research [4]. Another research have published about design and studying new technique called Support Vector Machine (SVM) based on experimental experience; Kazem et al obtained results proofed that the studied model achieved less MSE in comparing to the previous research about management energy generation [5]. Moreover; Salmi et al developed a new model photovoltaic cell by using Matlab/SIMULINK, the model is based on mathematical equations and described through an equivalent circuit including a photocurrent source. In addition an experimental test bench was built and the obtained results exhibited a good agreement with the simulation ones [6]. Regarding studying novel modeling technique of PV module with a fuzzy logic based MPPT algorithm and boost converter in Simulink environment which was developed by Mahamudul et al [7]. The main result has given a important contribution which is simplification of PV modeling technique and implementation of fuzzy based MPPT system to track maximum power efficiently. Other research about developing methodology for production of hydrogen by electrolysis system was proposed by Ayati et al [8]. However, in Terms of studying the over-current in solar systems; many researchers have been developed by using electrical relay. Barsoum et al [9] have studied overvoltage issue of electricity produced from renewable energy in power grid, the over-current relays due to reserve power to the 33 kV. The obtained results is illustrated that the transmission losses are reduced. Finally; Ok et al [10] have published a research about protection relay in wind power systems, where the over-current relay (OCR) based on wind speed. The benefits were verification of false actions by OCR's and the the electrical faults in the power grid can be solved.

Regarding the studying model; the Figure 1 shows the function of solar panel for producing hydrogen from $H_2SO_4$. However; the protection of the hybrid solar panel from over-current has taken the major part.

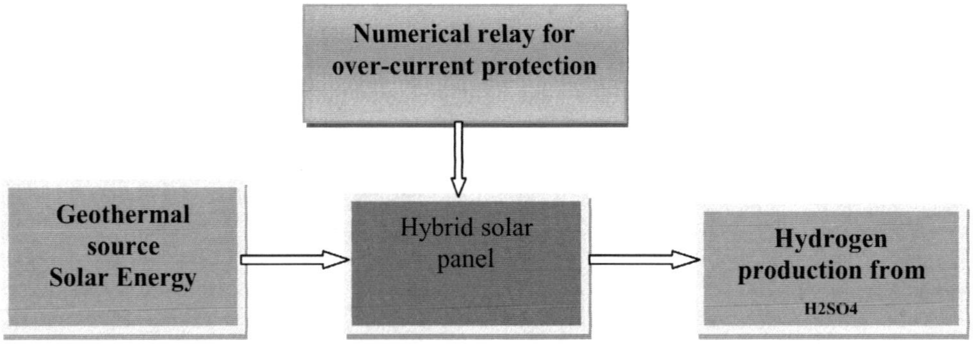


Fig. 1. Description of the studied case

The experience has been developed in Ghardia which is located in the south of Algeria due the presence of the solar panel station. For clarify the electricity capacity available in Algeria from the solar energy, the Figure 2 explains the solar electricity per area in Algeria [11], and the Figure 3 is area where the experience has been developed.

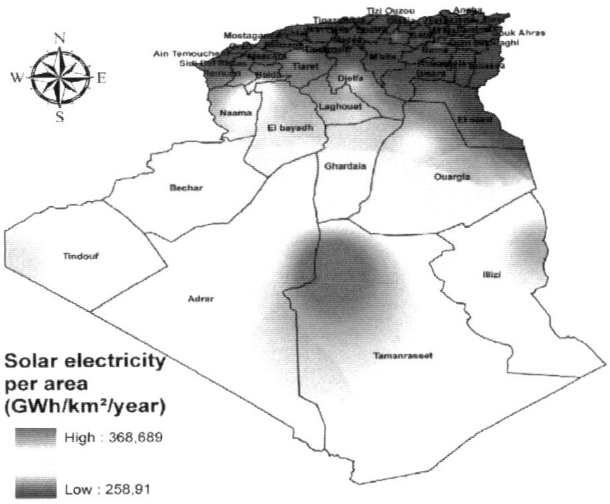


Fig. 2. Solar electricity per area in Algeria

Fig. 3. Solar Power Station in Ghardaia (Algeria).

## 2 IMPACT OF OVER-CURRENT ON PV SOLAR EFFICIENCY

Electrical Power System Protection is required for protection of both the user and the electrical system equipment from faults. It is important to validate the settings of power protection equipment and to confirm its performance when subject to different fault conditions. Consequently, the protective relay testing is very important before any new type of relay is commissioned in practical applications. Therefore; Numerical relays are currently being utilized for ensure various protection functions simultaneously. The over-current protection is the most extensively used to safeguard all power systems from the detrimental effects of faults causing high current flows. The fig 4 represents a real impact of over-current on solar system. As a result controlling the power station by installing the numerical relay REF542plus in Ghardia solar power station is very important to avoid technical problems in one side, and secure hydrogen production in the other side.

 https://doi.org/10.1051/e3sconf/20186100005

Fig. 4. Risk of Over-Current in PV Solar Panels [12].

# 3 PRACTICAL STUDY (OVER-CURRENT) IN SOLAR PANEL SYSTEM

Like any other device, the numerical over-current relay should be tested before its installation in the area to be protected. In this part, we will discuss one of the newest methods of testing which was realized in the laboratory.
The practical part consists testing a numerical Relay for protecting the solar panel from damage, and securing the continuation of the hydrogen production. So the passage from the first part which production of electricity from renewable energies to second part which hydrogen production has to pass this practical test.
In this part, we interfaced the relay under test to a PC through an optical connector placed in the front of the HMI. This interface permits to download the relay's configuration file from/to the PC. The base unit and the HMI of the relay are supplied with 100V dc and 48V dc respectively and are connected to each other using the physical interface RS485 standard connector. In the other side, we directly feed the CMC356 with 220V AC supply and linked to the relay using cables coming out of its three phase voltage and current outputs and going to the analog input module of the relay's base unit. This connection permits the CT and VT located inside the CMC356 to sense and to measure the injected voltages and currents before sending them to the relay. The communication between CMC356 and the PC where Test Universe is installed was done using an Ethernet cable. After finishing the connections above, the CMC356 and the relay were turned on, the relay was reset and so, ready to be tested. Figure 5 shows the REF452 plus and CMC356 connected together in the laboratory for tests.

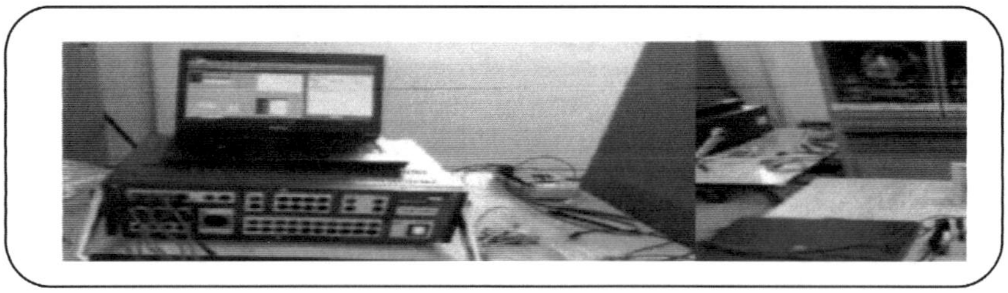

Fig 5. Studied Test of Over-Current in PV Solar panel.

# 4 PROSPECTS OF HYDROGEN PRODUCTION FROM HTES SYSTEM IN STABLE CASE AND INSTABLE CASE

## 4.1 Proposed Panel

Production of hydrogen from H2SO4 at 622°C in the studied case is possible by using the HTES technology where the temperature varied between 600°C and 1000°C. However there is another possibility to increase the temperature until 1000 from the same hybrid solar panel, and start producing hydrogen from different cycles. The Figure 6 shows the selected solar panel.

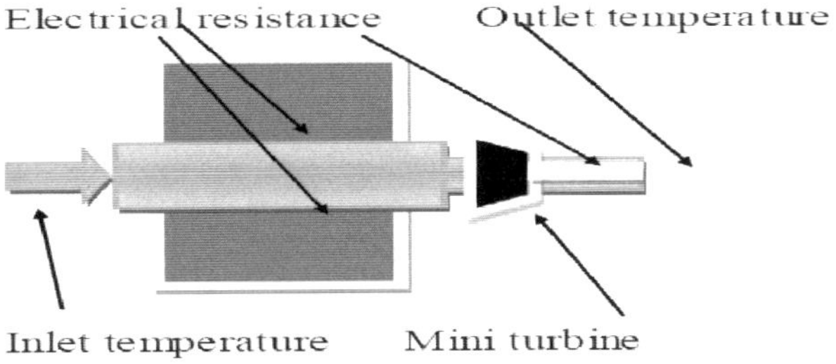


Fig. 6. Proposed PV Solar Panel

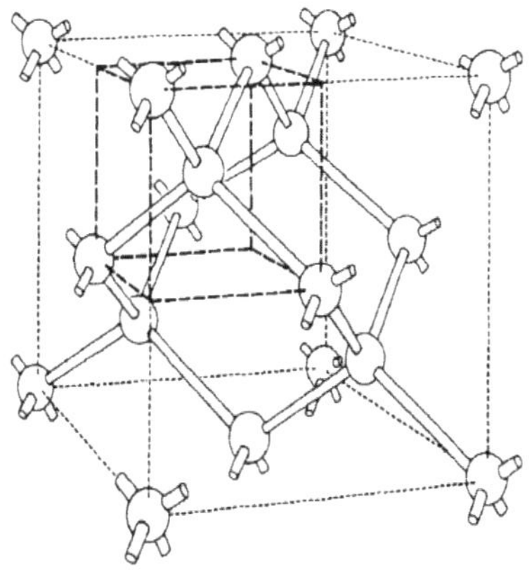

Fig. 7. Representation of the Silicon Crystal Lattice Arrangement [13]

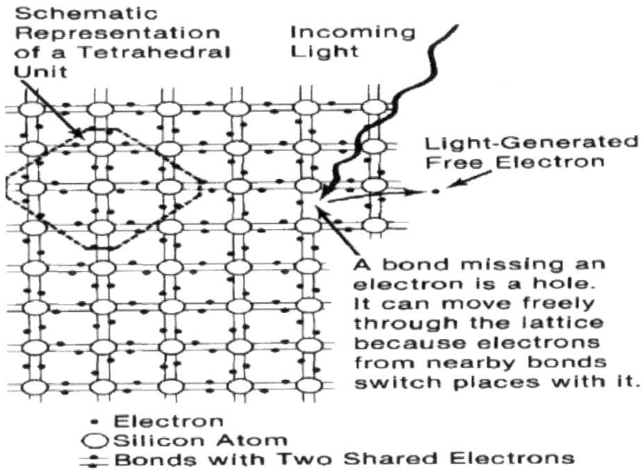


Fig. 8. Light Sufficient Energy Can Generate Electron Hole Pairs in Silicon [13]

In terms of hydrogen production from the selected solar panel; H2SO4 cycle has been selected for understanding the impact of the over-current on selected systems

H2SO4.

$$SO_2+I_2+2H_2O \rightarrow H_2SO_4+2HI \quad (100°C\text{-}130°C) \quad (1)$$

$$H_2SO_4 \rightarrow SO_2+H_2O+1/2\ O_2 \quad (800°C\text{-}1000°C) \quad (2)$$

$$2HI \rightarrow H_2+I_2 \quad (300°C\text{-}400°C) \quad (3)$$

Productivity of Hdrogen from $H_2SO_4$ is:

295.5 MJ/Kg$H_2$

## 5 RESULTS AND DISCUSSIONS

Controlling the over-current by installing Numerical Relay REF542plus is one of the most important factors for securing the hydrogen production and increasing the energy efficiency. The Table1 shows that amount of $H_2$ produced dropped by 89.537% in case of over-current uncontrolled (Absence of Numerical Relay REF542plus).

**Table 1.** Impact of Over-current on Hydrogen production from Hybrid Solar Panel from $H_2SO_4$.

| Conditions of Hydrogen production | Amount of hydrogen (Kg $H_2$) |
|---|---|
| 1/ PV solar panel under control (presence of Numerical Relay: REF542plus) | 0,517 Kg $H_2$/day |
| 2/ PV Solar panel Instable due to over-current (absence of Numerical Relay: REF542plus) | 0,05409 Kg $H_2$/day |

## 6 CONCLUSION

The aim of this project was to offer a good reference about the over-current protection. The Numerical Relay is the most appropriate for this type of protection, because it can accommodate different types of relay functions and provide an easy modification to tripping characteristics and change in conventional setting. Wrong settings in over-current relay parameters can lead to serious problems; therefore, testing it before any use is necessary and requires a full study to guarantee its efficient performance for all time. The test experiment for over-current function, performed on the relay confirms the correctness of its operation in accordance with manufacturer's design and suitability for intended application. The results were satisfactory and as expected.
The amount of hydrogen production in case of over-current none controlled is lower by 89.537% than the normal case in the proposed hybrid solar panel.

## References

1. J. F. Wilson , S. S. Srinivasan, B. M. Moore, L. Henderson, S. Ely , P.C. Sharma, Hydrogen Production Using Solar Energy, Journal of undergraduate research in physics 17/03/2013.
2. M. Douaka, N. Settoub, Estimation of hydrogen production using wind energy in Algeria, Energy Procedia 74, 981– 990, (2015)

3. V. Quaschning and F. Trieb, Solar Thermal Power Plants For Hydrogen Production, Hypothesis, IV Symposium Stralsund, Germany 9-14 September 2001, pp.198-202.
4. A. Acakpovi, E.B. Hagan, Novel Photovoltaic Module Modeling using Matlab/Simulink, International Journal of Computer Applications, 2013 by IJCA Journal
5. Al-Waeli, A. H. A., Sopian, K., Kazem, H. A., Yousif, J. H., Chaichan, M. T., Ibrahim, A., Ruslan, M. H., Comparison of prediction methods of PV/T nanofluid and nano PCM system using a measured dataset and artificial neural network. Solar Energy, 162, 378- 396, (2018).
6. T. Salmi, M. Bouzguenda, A. Gastli, A. Masmoudi, MATLAB/Simulink Based Modelling of Solar Photovoltaic Cell, International Journal of Renewable Energy Research, Vol.2, No.2, (2012).
7. H. Mahamudul, M. Saad, M. I. Henk, Photovoltaic System Modeling with Fuzzy Logic Based Maximum Power Point Tracking Algorithm, International Journal of Photoenergy Vol. , (2013).
8. F. Ayati, A. M'Raoui, M. Belhamel, A. Rebai, Modélisation d'un Système de Production d'Hydrogène Solaire par Electrolyse, Rev. Energ. Ren. Vol. 7, 135-150, (2004)
9. N. Barsoum, C. Z. Lee, Simulation of Power Flow and Protection of a Limited Bus Grid System with Injected Solar Power, Energy and Power Engineering, Vol. 5, 59-69, (2013).
10. Y. Ok, J. Lee, J. Choi, Analysis and Solution for Operations of Overcurrent Relay in Wind Power System Energies, Vol 9, 458, (2016).
11. S. Rahmouni, B. Negrou, N. Settou J. Dominguez, A. Gouareh, Prospects of hydrogen production potential from renewable resources in Algeria, International Journal of Hydrogen Energy Vol. 42, 1383-1395, (2017) .
12. https://www.solarpowerportal.co.uk/news/solar_pv_fires_predominantly_caused_ by_inst aller_error_government_report_co.
13. Technical information office, U.S Department of Energy, Basic Photovoltaic Principles and Methods, February 1982, SERI/SP-290-1448.

# An Investigation into the Effects of Risks and Risk Management of Bioenergy Projects

*Aminu* Bature[1*], *Lynsey* Melville[1], *Khondokar* Mizanur Rahman[1], *Jahangir* Akhtar[1], and *Poonam* Aulak[1]

[1]School of Engineering and the Built Environment, Birmingham City University, United Kingdom

**Abstract:** Risk is an important component in project management and plays a key role in securing project finance in renewable energy sector. As the amount of companies making investment in renewables grow, so do the inherent risks in the different stages of the projects. This study seeks to obtain information, which will help examine the risks associated with bioenergy projects and the methods used to identify, manage and reduce them. The study is conducted in the form of a survey with data being gathered using questionnaires. The results of this study shows that although when taken as a whole, policy and regulatory risk is the most significant risk associated with renewable energy projects, financial and environmental risks are the most important risks related to bioenergy projects. These suggest that financing, which is one of the key elements in maintaining, building and operating bioenergy plants faces substantial obstacles in managing it risks. Moreover, the study shows that insurance coverage and company-based risk management functions are the most important risk management tools for managing financial and environmental risks respectively. It is recommended that governments should strive to include bioenergy stakeholders in their policy development in order to integrate industry's perspective into the process.

**Keywords:** Bio-energy; Renewable Energy; Project Management; Risks, Risk Management

## 1 Introduction

According to the United Nations Environmental Programme (UNEP) and Bloomberg New Energy Finance (BNEF), the global investment in renewable energy was more than twice that of coal and gas-powered generation capacity in 2015 [1]. The amount invested rose to $285.9 billion, some 5% more than 2011's record of $278.5 billion. The use of bio-energy sources including biomass, biofuels and bio-power, have particularly increased by a factor of five from 2001 to 2011 due to efforts to reduce the environmental externalities of fossil fuels [2], policy incentives through support schemes, and the growing emphasis on energy security. These however, raises the investment risks in renewable energy projects, and the availability of risk management resources such as risk expertise, insurance cover and industry data to manage the risks remains limited. Thus, potentially restricting cash flows, which

* Corresponding Author: Aminu.Bature@mail.bcu.ac.uk

limits the industry's access to development capital. Although there has been an increasing interest in the literature on bio-energy technology for energy related greenhouse gas (GHG) emission mitigation [3; 4; 5], there is comparatively little discussion about the risks inherent in bioenergy projects and the approaches taken to manage them [6; 7].

This study seeks to obtain data that will help to examine the risks associated with bio-energy projects and the methods used to manage and reduce them. The study is conducted in the form of a survey (of over 50 renewable energy, and project and risk management experts), with data being gathered via questionnaires. The researchers also adopted the literature review approach to identify the significant risks commonly associated with bioenergy projects. The study offers some important insight into the most important risk(s) facing the bioenergy industry, the risk management challenges in the renewable energy sector, and the most common risk identification and transfer mechanism(s) used by experts in the industry.

The research shows that despite 59.65% of the respondents indicated that risk is "clearly defined and well communicated" in their organization, 40% of them also say that there is no standard risk management framework in their organization to implement risk management systematically and effectively. These suggest that the renewable energy sector faces obstacles in implementing risk management strategies. Moreover, financial risk is the most significant risk associated with bioenergy energy projects (flagged by 20.79%.), followed by environmental risks (flagged by 19.80%), operational (flagged by 16.83%) and political and regulatory risks (flagged by 12.87). However, 43.40% of the survey respondents say that political and regulatory risks are the most significant risks associated with renewable energy projects. These indicate that although overall politicy and regulatory risk(s) are considered high profile, they may not be the main risk factors in the bioenergy industry. Yet, consistency in regulation and policy may play an important role in how companies investing in the bioenergy industry perceive risks.

## 2 Materials and Methods

In order to investigate renewable energy project risks and the effect of risk management on bioenergy projects, the researchers first derive from the literature the different categories of risks, which formed the bases for this study's survey. A comparative approach is used to discuss the survey results concerning the relevant risks, citing peer-reviewed academic journals, company reports, and books. Although there is no universal agreements regarding the optimal length of a survey, simple and straight to the point questionnaires generally attract high response rates than very long complex ones [8]. Moreover, meticulous attention was paid to make sure that each question is appropriate, relevant, intelligible, precise, and free from any kind of bias. The questionnaire was sent to 65 professionals in the fields of renewable energy, risk management, and project management. The target respondents were selected based on their occupational titles (such as renewable energy experts, project and risks managers). The response rate was 92% (60 response), with 76% of the respondents from the private sector, 15% from the public sector and the remaining 9% from other sectors.

Quantitative studies such as this are subjected to certain degree of indeterminacy *vis-à-vis* their intrinsic strength to depict the surveyed phenomenon. Moreover, the respondent's ontological view about reality fundamentally makes the results of the survey for this study essentially subjective. Another limitation is the survey questions, which are open to interpretation between the researcher's epistemological view of the "truth" and the "real truth". Another limiting factor of this study is the relatively small total amount of responses to the questionnaire. All these limitations need to be considered when examining the survey results of this study. However, notwithstanding it is to be believed that these findings can be beneficial to the body of knowledge of renewable energy particularly bioenergy, and to the project and risk management community.

## 3 Results & Discussion

### 3.1 Risk Definition and Risk Management Framework in the Renewable Energy Sector.

Risk communication is an interactive process were stakeholders (such as institution, individuals and groups) exchange information and opinion about the nature of risks for risk management. It provides the means through which the public understand risk and defines the safety acceptance criteria [9]. According to the Institute of Risk Management (IRM), managing risk should be a continuous task that runs throughout an organization's strategy development and implementation [25]. Thus, understanding how well the risk is defined and communicated, and the standards or framework used is one of the key elements for understanding risk perception, risk management maturity and its impact in the renewable energy sector. When asked "*How well is risk defined and/or communicated in your organization?*" The majority of the respondents (59.65%) replied that risk is well-defined and communicated between risk assessors and risk managers in their organization, which indicates that the potential for effective risk assessments in decision-making within these companies is high. Moreover, 15.79% of the respondent expressed their belief that each project manager in their organization defines his/her own risk(s). Another 12.28% of the respondents answered that, there is no clear definition of risk in their organization, as shown in Figure 1. These suggest that the transfer of risk related information and their management as well as the exchange of knowledge and values as a preventive activity to cope with hazard events and reduce impact on people, are well established within the renewable energy sector. However, this does not mean that there is always agreement among different stakeholders.

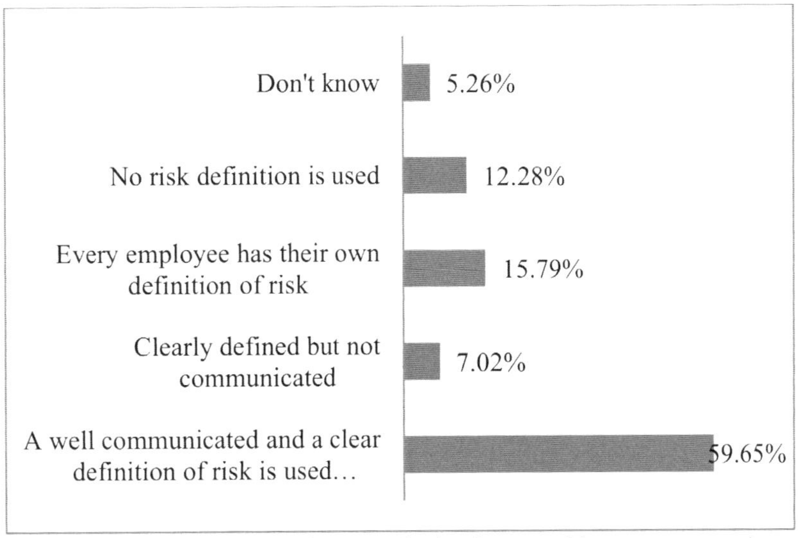


**Figure 1.** Risk definition and communication in renewable energy companies

Despite the unanimity of the respondents (59.65%) in this research and the corroboration from the IRM survey in 2008, the mere definition of risk and/or communication is not enough. It is important not only to define and communicate the risks, but also to have a guiding framework that meets international standard, such as IRM [10] and ISO 31000:2009 standard [11] for effective implementation. When asked if there is any risk management framework in used in their organization? 40% of those who responded indicated that there is no standard risk management framework in their organization. However, 29.09% of the participants responded that a "company owned or IRM standard enterprise wide frameworks" is used in their organization. 18.18% of those who responded answered that every department in their organization have their own risk management framework, and a small number of the respondents (12.73%) indicated that they do not know. These suggests that. Only a limited amount of renewable energy based companies have an established risk management framework. One respondent further states that "renewable energy project risks are often very complicated, confusing and lack consistency, therefore every department in his company has its own risk management framework. Therefore, most renewable energy consultants end up creating their own bespoke risks management methods and procedure".

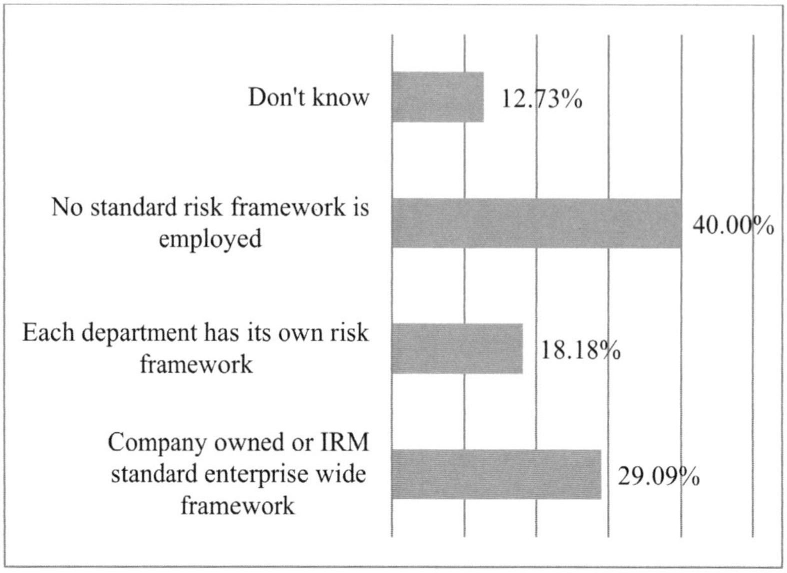


**Figure 2.** Risk Management Framework Adoption in the Renewable Energy Sector

This findings are inconsistent with those of previous work by the IRM, where more than half of their respondents (66.3%) indicates that enterprise-wide framework (such as IRM standard) promoted by risk management team are widely used in their organizations such as Banking and Finance, Construction and Property, Food and Beverage, Technology, Media and Telecoms *et cetra*.

## 3.2 Risk Identification, Assessment, and Response Action

### 3.2.1 Risk Identification

This is the first step in the risk management cycle [12] that "stems from the decision-making conditions under which an investor is at the moment" [13]. Risk identification has been defined as the process of identifying all the potential risks associated with the objectives of the project [14]. In the Project Management Body of Knowledge (PMBOK), the PMI refers risk identification as "the process of determining which risks may affect the project and documenting their characteristics" [15]. According to a definition provided by the OGC's PRINCE2, risk identification is the step "that obtains information about the project in order to understand the specific objectives that are at risk and to formulate an appropriate risk management approach," the result of which can be found in the risk register of the project. This step in the risk management cycle also ensures the incorporation of lesson learnt from previous projects [16]. Moreover, subsequent stages of the risk management cycle such as risk assessment and control will not be possible to undertake when/if the risks have not been identified [17].

In the last decade, various studies have attempted to identify the different types of risk associated with renewable energy projects. For example, Xu *et al.* [39] identified 5 Critical Risk Factors (CRFs) by conducting content analysis on the risk events data of 14 waste-biomass incineration generation projects in China. These factors are enironmental risk, payment risk, insufficient waste supply, lack of support infrastructure, and disposal of non-lincensed waste. However, Langholtz *et al.* [2] in their review of the evidence from the 2012 U.S. drought, identifies six risks in the bioenergy industry due to extreme weather events including shortages of biomass feedstock, higher feedstock cost, product price volatility, seasonal scarcity, input price volatility, and the availability of water. Lee and Zhong [38] introduced the hybrid bond project financing instrument, illustrating how the bond could be used to finance and manage major project uncertainties including quantifiable: credit risks and market risks, and unquantifiable: operational risks, liquidity risks, and political risks. To investigate both the internal and external risk factors in the process of implementing a renewable energy project, Rolik [40], employed strategic management tools such as SWOT analysis and McKinsey matrix to classify the risks into: market risks, project risks and management risks. This view is supported by Jinrong and Enyi [41 ]who writes that risks inherent in energy projects include market risk, project quality risk, management risk, as well as technology risk, , political and legal risk, client risk, and financial risk. However, in another major study, Guerrero-Liquet, *et al.* [43] investigated risk management tools in renewable energy facilities based on the Project Management Institute's (PMI) Project Management Body of Knowledge (PMBOK® Guide), using information gathering (Delphi and Checklist) and diagram techniques (Fishbone Diagram (Ishikawa)) to categorized risks into technical risks, price risks and financial risks. A broader perspective has been adopted by Gatzert and Kosub [35] who conducted an in-depth literature review on the different type of risks associated with renewable energy projects and categorises them into strategic/business risks, transport/construction/completion risks, operation/maintenance risks, liability/legal risks, market/sales risks, counterparty risks, and political/policy/regulatory risks. Together these studies provide important insights into the different risks asscoiated with renewable energy projects. Moreover, the studies presented thus far provide evidence that suggests that by far the greatest risks associated with renewable energy projects including bioenergy projects may be categorised into *technical risks, political and regulatory risks, financial risks, social risks, and environmental risks*.

In this study, 94.8% of those who responded to this question were found to use some type of risk identification technique or approach to identify risks, with the exception of 5.2 % as shown in Figure 2. This finding is in agreement with the Economist Intelligence Unit (2011) findings, which showed that (70%) companies in the renewable energy sector are highly successful in identifying risk [18]. Moreover, this result corroborate the findings of previous work by the IRM on risk in other sectors. There are several techniques available to identify risks depending on the information and project players in hand. In the questionnaire developed for this study, the researchers present the participants with a set of risk identification techniques (see Figure 3 below) commonly found in the literature and used by practitioners [19; 20; 17].

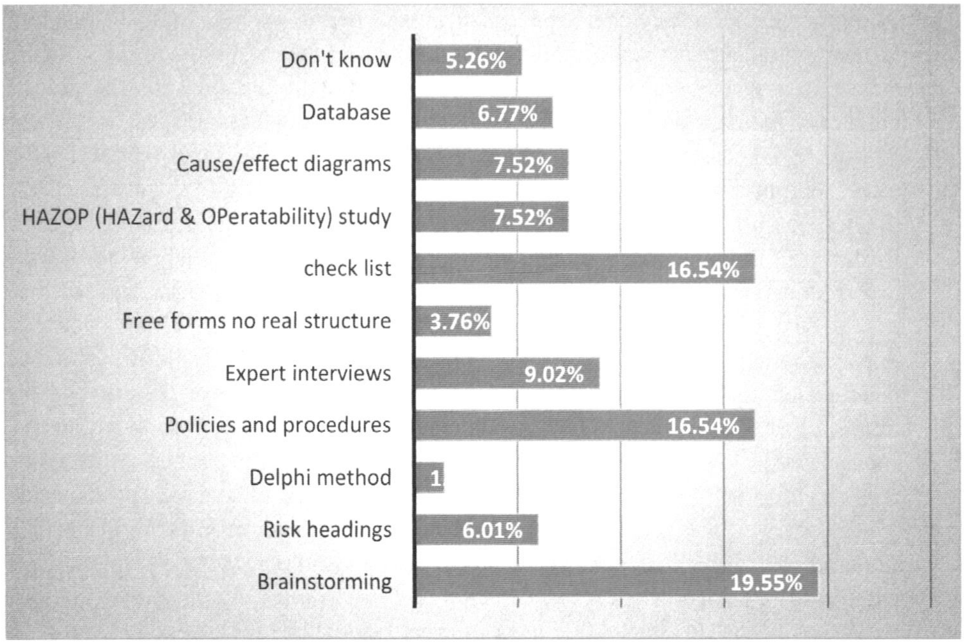


**Figure 3.** Risk Identification Techniques used in the Renewable Energy Sectors

### 3.2.1.1 Risk Identification Techniques

#### 3.2.1.1.1 Delphi Technique

Dalkey and Helmer of RAND Corporation in California first developed this technique in the 1950s for gaining convergence of opinions concerning real-world knowledge solicited from subject matter experts within a given topic areas. Ever since, this method has been used by both practitioners and academics to collect data from respondents within their topic expertise [20]. It involves creating ways to gain agreement or disagreement (consensus) among experts in relation to a given problem by expressing their honest opinions about the risk posing threats to the project [22]. According to the IEA, the Delphi method also allow experts to revise their opinions and come up with new ideas [21]. Having discussed what the Delphi method is, this section of the questionnaire is designed to investigate common methods of identifying risks in renewable energy projects. Since more than one method can be used for large and complex projects with multiple departments, the respondents were offered a "select all that apply" response option. When asked which risk identification technique(s) do you used in your organization? Only 1.50% of the respondents choose this method.

#### 3.2.1.1.2 Brainstorming

This "is a decision-making technique used to indicate verbal generation of ideas by a group" [23]. The technique was first introduced in the 1930s by Alex Osborne to generate and clarify ideas of potential risks by bringing together project stakeholders and subject matter experts under a facilitator [19; 21]. Osborn [45] defined brainstorming session as an approach that utilizes group of people to generate many ideas. According to Forbes *et al.* [17], brainstorming technique tend to be presented in a structured or unstructured (simple) approach. The objective of this risk identification technique is to collect as much ideas from a group of people as possible within a given period of time. The session begins by making sure that the participants understood very clearly the objectives of the gathering and that they are there to collaborate with each other. The goal of the session is to identify risks that are known unknowns as well as those that were unknown unknowns. From the survey, brainstorming is the most common method of identifying risk (19.55%). This result is broadly consistent with earlier research [24; 25].

#### 3.2.1.1.3 Expert interviews

This is considered the simplest form of risk identification were a series of questions are presented to industry experts to obtain their opinion in connection to potential risks [21]. This technique is widely used in different stages of the projects, and may be used to cover both internal and external risks [26]. It can be unstructured, semi structured, or structured [27]. Of the 54 participants who responded to this question, 9.02% reported to use expert interviews to identify risks. This is in contrast to earlier findings by Rostami [26], who found that expert judgement is the second most used risk identification technique in the construction phase of projects with a 71% usage rate.

#### 3.2.1.1.4 HAZOP (HAZard and OPerability analysis) and Other Risk Identification Techniques

HAZOP is the process of identifying project hazards due to operational setbacks in the project activities [21]. Only 7.25% of those who responded indicated that HAZOP is used in their projects. According to the literature, other key techniques in identifying risks are: check-list, cause and effect diagram, data base (such as lesson learn logs and risks registers) and risk headings. Check list, and policies and procedures both 16.54%. Some respondents indicated that cause and effect diagrams (7.52%) is used in their organizations. A minority of participants (6.77%) uses database or company documentation review method, which differ from Rostami [26] result were 75% of the respondents considered this technique as their most preferred tool for identifying risk. (6.01%) uses risk headings to identify risk in their projects. Moreover, three participants offers other methods used in their organizations such as internally development ERM matrix, probability distributions, and successive principle by Steen Lichtenberg.

### 3.2.2 Risk Analysis and Assessment

The term risk analysis is open to a series of definitions. For example, Moddares, *et al.* [28], defines risk analysis as a method for identifying, characterizing, quantifying, and evaluating hazards. Martinez [29] described risk analysis as a proactive type of methodology for the identification of weaknesses and threats to organization's assets. A further definition is given by Galway [46] who describes risk analysis as the process of qualitatively and quantitatively assessing risk. In the field of renewable energy, Ferraris *et al.* [47] defines risk analysis as the process of scientifically characterizing "what is known and what is uncertain about the present and future performance of the system under examination". In this research, the term risk analysis will be used interchangeably with (or to include) risk assessment when referring to all measures taken to determine the likelihood of failure in meeting project objectives; to improve how decisions are made within the project, and; to help minimize the project exposure to risks [28; 29; 30]. Some of the methods used in risk analysis include the stochastic critical path method (CRM) for schedule risks, and the stochastic simulation of cost, both of which require Monte Carlo simulation and specification of uncertainty for the project time and cost. According to Razaque *et al.* [48], this stage in the risk management cycle helps reduce risk and anticipate the future repercussions of decision taken. Moreover, it involves the identification of the risk control mechanisms, and the resultant amount of controlled and residual risks [12; 31].

The literature outlines two methods for this stage namely qualitative approach such as Analytical Hierarchy Process (AHP) [32; 33] and risk matrix, and quantitative approach such as Fuzzy Logic, Failure Mode and Effect Analysis (FMEA) Tree, and Event Tree Analysis (ETA). The qualitative analysis phase focused on "the correct estimation of project's risk probability and scale accompanied effects" by using number tools while the quantitative analysis "presents tangible benefits" of what has been analyzed [34; 13]. According to Berg [12], qualitative or semi-quantitative techniques such as risk graph, hazard matrices, risk matrices (most common) and monograph are usually used for screening lower risks, whereas, quantitative techniques are used for screening higher risks. In order to assess the significant of the different categories of risks identified in the literature, the participant were asked to rate them into high, medium and low on the risk scale. Table 1 below shows the result of the survey for this study.

**Table 1.** Risk Scale for Assessing Renewable Energy Risks

| Risk category | Low | Medium | High |
|---|---|---|---|
| Political/Regulatory risk | 28.30% | 28.30% | 43.40% |
| Financial risk | 11.32% | 50.94% | 37.74% |
| Technology risk | 47.17% | 35.85% | 16.98% |
| Operational risk | 42.59% | 38.89% | 18.52% |
| Environmental risk | 64.15% | 13.21% | 22.64% |
| Construction risk | 42.59% | 38.89% | 18.52% |
| Weather related risk | 20.75% | 54.72% | 24.53% |
| Social risk | 72.55% | 17.65% | 9.80% |
| Other risks | 67.44% | 23.26% | 9.30% |

Nearly half of the respondents (43%) agreed that political and regulatory risk have high impact on their projects. This result is consistent with previous research findings such as [35; 18; 36; 37]. For instance, Skytte *et al.* [36] uses qualitative survey to quantify the challenges that arise from change in political support systems and risk with respect to investment renewable energy technologies. The authors found that "political risk factors are the most" important and "difficult ones to include in the investment considerations". That is, very often risk planners "do not include (political risk factors) when calculating a risk premium". Cleijne and Ruijgrok [50] aimed to develop a model to describe and quantify the dynamics of renewable energy implementation in Europe focusing on risk as perceived by the stakeholders. The responses to their empirical data also suggest that regulatory/political risk on tariffs/financial support have high impact and are regarded to as important factors that are difficult to predict. In the same vein, Komendantova *et al.* [37] found that stakeholders in renewables in North Africa consider policy and regulatory risks as critical risk. Moreover, in corroboration with the finding of this study, James Green head of renewable energy practice at JLT Specialty in 2011 observed, "In the UK I was looking at 40 or so 5-mw ground-mounted solar installations, which we were asked to insure. And overnight when the UK government said that it was going to cut the feed-in tariff, the 40 installations literally went down to four" [18].

The majority of those who responded to this survey felt that financial and weather related risk have medium impact on their projects. Although, these results are consistent with some studies, for instance [18], they differ from those conducted by Apak et al. [49] and the IEA [21]. These studies found that financial risks have high impact on renewable energy plants return on equity and on debt leverage capability and is usually included in probabilistic Discounted Cash Flow (DFC). Moreover, when the participants were asked to rate the impact of technological, operational, environmental, construction and social risks, majority of the respondents indicated that these risk factors have medium impact on their energy projects. However, when the respondents were asked the same question in relation to bio-energy projects specifically, the first risk category to emerge was financial risks with 20.79% of the respondents indicating that financial related risks have high impact on bio-energy projects. This finding is supported by Xu *et al.* [39], who writes that "payment of waste disposal fees and electricity fees require the support and cooperation of the government". The second category of risk to emerge was environmental risks as shown in Figure 4 below.

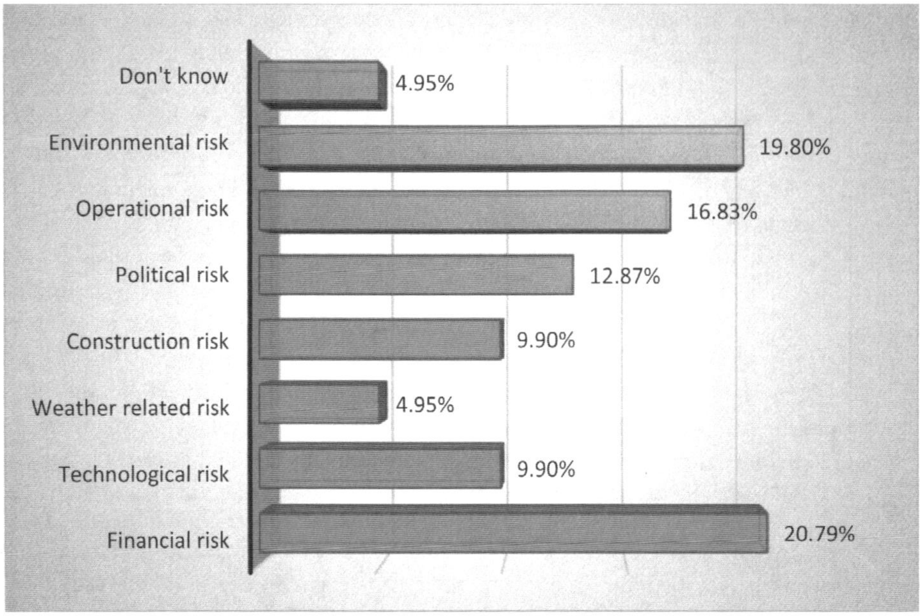


**Figure 4.** Risk assessment of Bioenergy Projects.

However, the findings of the current study do not support the previous research into the risks affecting the biofuel industry by Pries *et al.* [7]. The authors found that risk related to management and management process as well as profitability and market conditions were the most important risk affecting the biofuel industry.

### 3.2.3 Risk Response: Reduction & Mitigation

While some authors have mainly been interested in questions concerning risk identification and assessment [38; 39; 40; 41], Other are more concerned with risks response strategy and implemention such as risk acceptance [42], risk reduction [43], risk transfer *et cetra*. According to the Project Management Institute's PMBOK [15], risk mitigation is the effort of minimizing the likelihood, and/or consequences of an already identified risk in the project, prior to the occurrence of the risk event. According to a definition provided by the IEA [21], risk mitigation is "the risk handling strategy used for minimizing the probability of occurrence, negative impact of identified risk, increasing the benefits from positive risk (opportunity). These are actions taken to reduce the likelihood of an uncertain event as well as overcome the effects of risk. Risk mitigation principle may be presented at each phase of the project, from the initiation and planning stages to the closing a project stage [13]. Hitzeroth and Megerle [42] points out that attitude parameters adopted by renewable energy project stakeholders "which especially at the outset of such planning processes seemingly tend towards "acceptance" such as "incertitude" or "conditional acceptance" represent above-average pronounced risk factors in view of the project proponent's planning and cost certainty". Since project implementation (including bio-energy projects), depends on the effectiveness of the risk response mechanism adopted by both public and private stakeholders [39], it important to investigate the different types of response mechanisms used by practitioners in the renewable energy sector. Therefore, the participants were asked to select which risk response mechanism do they used to manage the various categories of risks.

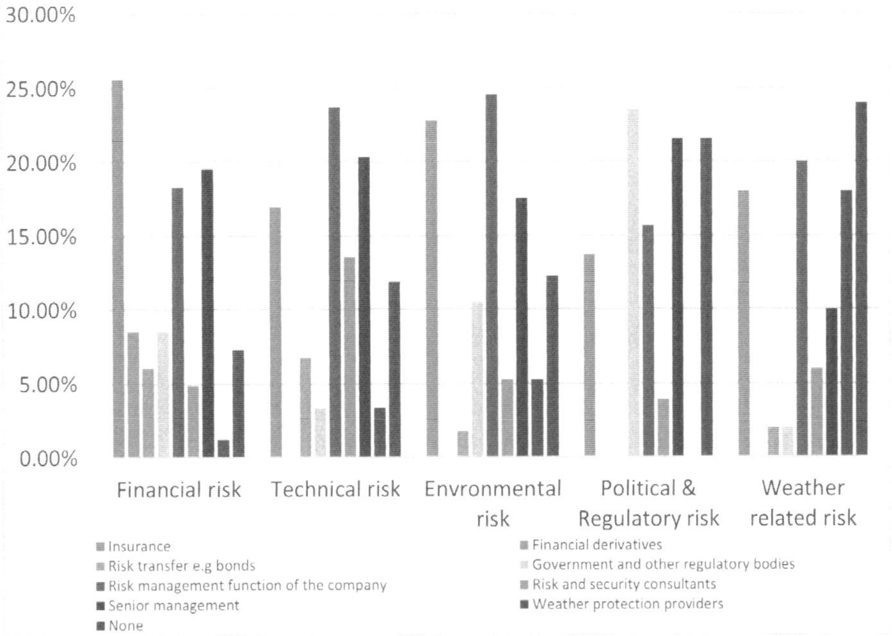


**Figure 5.** Risk Response Mechanisms for Different Risk Categories in Renewable Energy Projects

The majority of those who responded to the survey indicate that insurance, internal risk management function, government and other regulatory bodies, and none (that is no risk response) are the most common risk response mechanisms used to manage/response to financial, technical and environmental, political, and weather related risks respectively. Of the 60 practitioners who completed the questionnaire, 24% indicated that they do not have any risk response mechanisms for managing weather related risks. Only a small number of the respondent indicated that financial derivatives (such as bonds, market indexes and stocks) were used to manage financial risks. Overall, these results provide important insight into significant of buying insurance to manage risks in the energy sector. As the head of the insurance service group of the international finance corporation, Jan Mumenthaler put it: renewable energy developers should stop seeking to manage risks through risk transfer means, but should buy insurance to deal with it (risks). Xu *et al.* [39] also came to a similar conclusion about financial and environmental risk response in their study on the "critical risk factors affecting the implementation of PPP waste-to-energy projects in china" were they suggested that such risks are "better handled by the private entity" such as insurance company. Moreover, Gatzert and Kosub's [35] comprehensive study on current risks and risk management solutions of renewable energy projects shows that, insurance and diversification are the most important risk management tools used in renewable energy projects.

## 4 Conclusions

As renewable energy projects becomes more complex, so does the risks. The present study is designed to determine the effect of risks and risk management for bioenergy projects. According to the literature, increasing amount of companies in the energy sector are including renewable energy in their business strategies. The researchers identified five risks categories from the literature related to renewable energy projects, namely; technical risks, political and regulatory risks, financial risks, social risks, and environmental risks. According to the data collected in the questionnaire surveys, the most important risks are related to finance, policy and regulations, and the environment, which are intensified by the gloomy behaviour of the economy in many countries. Bioenergy companies consider weather related risks as less significant, despite been very important to other renewables such as wind and solar. Insurance is the main form of risk response mechanism used by companies in the bioenergy sector. Therefore, it is important to develop new financial coverages that may allow further growth in bioenergy's project financing. Moreover, political and regulatory risks were found to be managed mostly by government and other regulatory bodies external to the industry. Thus, political and regulatory risks will remain relevant to investors until policy makers begin to address these risks by integrating bioenergy industry stakeholder's perspective into their policy development process.

## Acknowledgments

The authors wish to acknowledge the petroleum technology development funds (PTDF), for their immense support and sponsorship for the study as well as the School of Engineering, and the Built Environment, Faculty of Computing, Engineering and the Built Environment, Birmingham City University, United Kingdom for their assistance.

## References

1. UNEP, 2016. *Global Trends in Renewable Energy Investment,* Frankfurt : Frankfurt School of Finance & Managemen.
2. Langholtz, M. et al., 2014. Climate risk management for the US cellulosic biofuels supply chain. *Climate Risk Management,* Volume 3, pp. 96-115.
3. Adler, P. R., Grosso, S. J. D. & Parton, W. J., 2007. Life-cycle assessment of net greenhouse-gas flux for bioenergy cropping systems. *Ecological Application,* Volume 17, pp. 675-691.
4. Schneider, U. A. & McCarl, B. A., 2003. Economic Potential of Biomass Based Fuels for Greenhouse Gas Emission Mitigation. *Environmental and Resource Economics,* 24(4), p. 291–312.
5. Wihersaari, M., 2005. *Aspects on bioenergy as a technical measure to reduce energy related greenhouse gas emissions.* Konetekniikan osasto: VTT Technical Research Centre of Finland.

6. Wu, Y. et al., 2018. A risk assessment framework of PPP waste-to-energy incineration projects in China under 2-dimension linguistic environment. *Journal of Cleaner Production,* Volume 183, pp. 602-617.
7. Pries, F., Talebi, A., Schillo, R. S. & Lemay, M. A., 2016. Risks affecting the biofuels industry: A US and Canadian company perspective. *Energy Policy,* Volume 97, pp. 93-101.
8. Leung, W., 2001. *How to Design a Questionnaire,* s.l.: STUDENT BMJ .
9. Goswami , D. Y. & Kreith, F., 2007. *Handbook of Energy Efficiency and Renewable Energy.* 1st ed. Boca Raton: CRC Press.
10. IRM, A., 2002. *Risk Management Standard.,* London: The Institute of Risk Management.
11. ISO, I., 2009. *31000: 2009 Risk Management—Principles and Guidelines,* Geneva, Switzerland: nternational Organization for Standardization.
12. Berg, H. P., 2010. Risk Management: Procedures, Methods and Experiences. *Reliability: Theory & Applications,* 2(17).
13. Szymański, P., 2017. Risk management in construction projects. Poznań, ELSEVIER Procedia Engineering, pp. 174-182.
14. Curtis, P., Carey, M. & Deloitte & Touche LLP, 2012. *Risk Assessment in Practice,* Durham: COSO: Commitee of Sponsoring Organizations of the Treadway Commission .
15. PMI, 2013. *A guide to the Project Management Body of Knowledge (PMBOK Guides).* 5th ed. Pennsylvania: Project Management Institute.
16. PRINCE2, 2017. *Managing successful projects with PRINCE2.* 6th ed. Norwich: The Stationery Office (TSO).
17. Forbes, D. R., Smith, S. D. & Horner, R. M. W., 2008. *A comparison of techniques for identifying risks in sustainability assessment of housing.* s.l., In Procs 24th Annual ARCOM Conference, pp. 1-3.
18. The Economist Group, 2011. *Managing the Risk in Renewable Energy,* s.l.: The Economist Intelligence Unit.
19. Balackova, H., 2002. *Brainstorming: A Creative Problem-Solving Method,* Prague: Masaryk Institute of Advanced Studies, Czech Technical University .
20. Hsu, C. C. & Sandford, B. A., 2007. The Delphi technique: making sense of consensus. *Practical assessment, research & evaluation,* 12(10), pp. 1-8.
21. IEA, 2011. *Risk Quantification and Risk Management in Renewable Energy Projects,* s.l.: Altran GmbH & Co., KG, IEA-RETD .
22. Cuhls, K., 2002. *Delphi Method,* Germany: Fraunhofer Institute for Systems and Innovation Research .
23. Brahm, C. & Kleiner, B. H., 1996. Advantages and disadvantages of group decision-making approaches. Team Performance Management. *An International Journa,* 2(1), pp. 30-35.
24. Wood, G. D. & Ellis, R. C. T., 2003. Risk management practices of leading UK cost consultants. *Engineering, Construction and Architectural Management,* 10(4), pp. 254-62.
25. IRM, 2008. *Innovation, Value Creation and Opportunity Special Interest Group,* s.l.: The Institute of Risk Management .
26. Rostami, A., 2016. Tools and techniques in Risk identification: a research within SMEs in the UK construction industry. *Universal journal of management,* 4(4), pp. 203-210.
27. Kansal, R. K. & Sharma, M., 2012. Risk Assessment Methods and Application in the Construction Project. *nternational Journal of Modern Engineering Research (IJMER),* May-June, Volume II , pp. 1081-1085.
28. Moddares, M., Kaminskiy, M. & Krivtsov, V., 1999. *Reliability Engineering and Risk Analysis: A Practical guide.* New York: Marcel Dekker, Inc.

29. Martinez, J., 2001 . *Integrated Risk Management: A Concept for Risk Containment.* s.l.:CRC Press LLC.
30. Merna, T. & Al-Thani, F., 2008. *Corporate Risk Management.* West Sussex : John Wiley & Sons.
31. Bowles, D. S., 2010. Dam Safety Risk Management for Hydroelectric Projects. *In International Conference on Hydropower Projects in Portugal: A New Cycle, Porto,* Volume 17, p. 2015.
32. Triantaphyllou, E. & Mann, S. H., 1995. Using the Analytical hierarchy Process for Decision making in Engineering Applications: Some Challenges. *International Journal of Industrial Engineering: Applications and Practice,* 2(1), pp. 35-44.
33. Saaty, T. L., 2008. Decision making with the analytic hierarchy process. *Int. J. Services Sciences,* Volume 1, pp. 83-98 .
34. APM, 2000. *Project Risk Analysis and Management* , Buckinghamshare: The Association for Project Management.
35. Gatzert, N. & Kosub, T., 2016. Risks and risk management of renewable energy projects: The case of onshore and offshore wind parks. *Renewable and Sustainable Energy Reviews,* Volume 60, pp. 982-998.
36. Skytte, K. et al., 2003. *Challeneges for Investment in Renewable Electricity in the European Union,* Petten: Alener & ADMIRE REBUS.
37. Komendantova, N., Patt, A., Barras, L. & Battaglini, A., 2012. Perception of Risk in Renewable Enerfy Projects:The Case of Concentrated Solar Power in North Africa. *Energy policy,* Volume 40, pp. 103-109.
38. Lee, C. W. & Zhong, J., 2015. Financing and risk management of renewable energy projects with a hybrid bond. *Renewable Energy,* Volume 75, pp. 779-787.
39. Xu, Y. et al., 2015. Critical risk factors affecting the implementation of PPP waste-to-energy projects in China. *Applied energy,* Volume 158, pp. 403-411.
40. Rolik, Y., 2017. Risk Management in Implementing Wind Energy Project. *Procedia Engineering,* Volume 178, pp. 278-288.
41. Jinrong, H. & Enyi, Z., 2011. Engineering risk management planning in energy performance contracting in China. *Systems Engineering Procedia,* Volume 1, pp. 195-205.
42. Hitzeroth, M. & Megerle, A., 2013. Renewable energy projects: acceptance risks and their management. *Renewable and Sustainable Energy Reviews,* Volume 27, pp. 576-584.
43. Guerrero-Liquet, G. C. et al., 2016. Decision-making for risk management in sustainable renewable energy facilities: A case study in the Dominican republic. *Sustainability,* 8(5), p. 455.
44. Bratt, G., 2010. *Study of Renewable Energy Project Risk Factors Influencing the Insurance Industry,* Glasgow: University of Strathclyde: Engineering .
45. Osborn, A. F., 2001. *Applied Imagination.* 3rd ed. New York: Creative Education Foundation Press.
46. Galway, L., 2004. *Quantitative Risk Analysis for Project Management: A Critical Review,* s.l.: RAND Coporation .
47. Ferraris, I., de la Canal, M. D. & Labriola, C., 2013. Risk analysis in renewable energy: assessment of the vulnerability of the environment and community. *Journal of Energy and Power Engineering,,* 7(8).
48. Razaque, A., Bach, C., Salama, N. & Alotaibi, A., 2012. Fostering Project Scheduling and Controlling Risk Management. *International Journal of Business and Social Science,* 3(14).
49. Apak, S., Atay, E. & Tuncer, G., 2011. Financial risk management in renewable energy sector: Comparative analysis between the European Union and Turkey. *Procedia-Social and Behavioral Sciences,* Volume 24, pp. 935-945.
50. Cleijne, H. & Ruijgrok, W., 2004. *MODELLING RISKS OF RENEWABLE ENERGY INVESTMENTS,* Austria: European Communities.

https://doi.org/10.1051/e3sconf/20186100006

51. OED, 2012. *Oxford English Dictionary.* 7th ed. Oxford: Franklin Watts.
52. Kytle, B. & Ruggie, J. G., 2005. *Corporate Social Responsibility as Risk Management: A Model for Multinationals,* Cambridge: Harvard Kennedy School.
53. Gaurav, S., Chileshe, N. & Ma, T., 2011. *Project risk analysis of solar energy project delays in India.* s.l., Solar2011, the 49th AuSES Annual Conference 30th November -2nd December .
54. Lee, P., Lam, P. T. I. & Lee, W. L., 2015. Risks in Energy Performance Contracting (EPC) projects. *Energy and Buildings,* Volume 92, pp. 116-127.
55. Michelez, J. et al., 2010. *Risk quantification and risk management in renewable energy projects,* s.l.: Altran GmbH & Co. KG or IEA - RETD.
56. Karytsas, S. & Choropanitis , I., 2017. Barriers against and actions towards renewable energy technologies diffusion: A Principal Component Analysis for residential ground source heat pump (GSHP) systems. *Renewable and Sustainable Energy Reviews,* Volume 78, pp. 252-271..
57. Papaioannou, M. G., 2006. *Exchange rate risk measurement and management: Issues and approaches for firms,* (No. 2006-2255): International Monetary Fund (IMF).

# Electronic Differential And Neuro-Fuzzy Sliding Mode Control With Extended State Observer For An Electric Vehicle System

*Ibrahim Farouk* Bouguenna[1,*], *Ahmed* Azaiz[1] , *Ahmed* Tahour[2] , *Ahmed* Larbaoui[2]

[1]RCAM Laboratory, University of Sidi Bel-Abbes, 22000, Algeria
[2]University of Mascara, 29000, Algeria

**Abstract.** In this paper a neuro-fuzzy-sliding mode control (NFSMC) with extended state observer (ESO) technique; is designed to guarantee the traction of an electric vehicle with two distinct permanent magnet synchronous motor (PMSM). Each PMSM systems ( source-converter-motor) are attached to an electronic differential (ED),in order to adjust the senses of direction of the vehicle ,and sustain a stable speed by adapting the difference in velocity of each motor-wheel according to the direction in the case of a turn. Two types of controllers are employed by a hybrid control scheme to assure the control and the performance of the vehicle. This hybrid control scheme guarantees the stability of the vehicle by ED, reduces the chattering phenomena in the PMSM electric motor, and improves the disturbance rejection ability which employs tow types of controllers. The neuro-fuzzy sliding mode control on the direct current loop and ESO controller on the speed loop, and the quadratic current loop; taking into account the dynamic of the vehicle. Simulation runs under Matlab/Simulink to assess the efficiency ,and strength of the recommended control method on the closed loop system.

## 1 Introduction

The intense use of electric car in the world has played a very important role in increasing the purity of the air in overcrowded areas. Limiting the impact of air polluted by the means of transport on humanity ,and has decreased the global warming of the earth [1].

To achieve cleaner and greener mobility in highly urbanized cities around the world, the utilization of electric vehicle must be considered as one of the solutions [2]. Many researches on electric vehicles has been carried out such as control system which is a very important part of the vehicle or electric vehicle's brain. It determines the total performance of the vehicle , as it is the key to improve motor efficiency. in our work a permanent magnet synchronous motor has been adopted as the electric vehicle (EV) propulsion.

Many researches has been done on electric vehicles testing the control of its traction chain, which is translated into the control of these engines wheels. As a result, The performance-dependent process of the engine-wheel assembly system is distinguished by

* Corresponding author: faroukspvusto@yahoo.fr

the smooth spin throughout the engine speed orbit, total torque control at zero speed, and rapid acceleration and deceleration.

On a standard vehicle the drive wheels are joined by half-shafts restrained by a bevel gear. On each of the wheels there is a torque equal to half of that provided by the return angle. Because of the hard connection arranged with the drive shaft, the two drive wheels revolve at the same angular speed, which has no specific disadvantage in walking in a straight line [3]. The track of the inner wheel is decreased contrasted to that of the outer wheel; For these two distances to be realized in the same time interval, it is essential that the angular speed of the two wheels is diverse. If the wheels are not driving, no problem, besides it is vital to insert a differential mechanism permitting the wheels to spin at dissimilar speeds. There is a system in classic vehicles that assures the difference in speeds while ensuring torque for the two driving wheels [4]; named mechanical differential. In this work we will substitute the mechanical differential with an electronic differential in order to lessen the drive line components, therefore enhancing the general dependability and efficiency. This option will also diminish the drive line weight since mechanical differential and gear reduction are not used. But, one of the main problems in the construction of these EVs (without mechanical differential) is to assure the vehicle steadiness specially while cornering or under slippery road conditions [5] . This calls for a specific traction control system . The control of the traction effort transmitted by each wheel is at the base of the command strategies intending to enhance the steadfastness of the vehicle. Where the wheels are controlled independently by using an electric motorization.

With in particular context the electric vehicle (EV) propulsion has been adopted by permanent magnet synchronous motor. These motors (PMSMs) have comprehensive industrial applications containing electric vehicles, robotics, and wind energy conversion systems as cited in [6].

This can be imputed to desirable performance characteristics comprising high power density, high torque to weight ratio, high reliability and high efficiency [6]. Many control techniques for PMSMs have been developed in literature studies. Although its poor performance the PI controller remains a popular choice because of its simplicity ,and ease of implementation, it does not take disturbances and uncertainties into account, leading to poor performance [7].

In consequence of the improvement that have been made to develop the digital signal processor technology, power electronics, control theories, more and more advanced control methods that are introduced to the PMSM control problem, e.g., adaptive control [8], robust control [9], sliding mode control [10], neural network control [11], and fuzzy control [12], etc. Which can ameliorate the electric vehicle's performance founded on the PMSM from different aspects.

Due to its ability to reject internal parameter variations and external disturbances, the sliding mode control is a popular control. SMC has found extensive applications in the areas of power electronics and electric machines [13]. As a result , SMC has been widely and successfully applied into the position and velocity control of PMSM. However, SMC like anything else has its own negatives, i.e., chattering phenomenon, which has originated from the interaction between parasitic dynamics and high frequency switching control [14]. For that, the control method was proposed to ensure the traction of the electric vehicle in particular to eliminate the chattering phenomenon , which is the neuro-fuzzy-sliding mode control (RNFSMC) with extended state observer (ESO). The hybrid neuro-fuzzy sliding mode control with extended state observer [15] is one of the most common controls that has associated several techniques: fuzzy control [16] and sliding mode control [17], neural network , and disturbances rejection laws to employ the advantages of all the techniques to restrain the disadvantages of regulation by conventional adjustment algorithms and to improve the performance of the controller system (stability, accuracy, speed, robustness, etc...).

The FIS forms are a useful computing framework based on the concepts of fuzzy set theory, fuzzy if–then rules and fuzzy reasoning. The ANFIS [21] is a FIS implemented in the framework of an adaptive fuzzy neural network. Combining the explicit knowledge representation of a FIS with the learning power of ANNs. Usually, the transformation of human knowledge into a fuzzy system (in the form of rules and membership functions) does not give the target response accurately. So, the parameters of the FIS should be determined optimally [24].

The lumped disturbances of the system which are composed of internal dynamics and external disturbances are considered as new state of the system by Extended State Observer (ESO) [18]. Where both states and lumped disturbances can be estimated by this observer. Based on the ESO, a feed forward compensation for the disturbances can also be employed in the control design.

In this paper, we try to explain the development of a robust control of an electric vehicle system, that helps to guarantee a stability and performances of the vehicle using ED and adaptating the closed-loop combining the extended state observer (ESO) with the adaptive neuro-fuzzy inference system technique together.

The content of this paper is arranged as follows: section 2 reviews the principle components of the traction system and their model equations (vehicle dynamic, Electronic differential , electric motor model). Section 3 demonstrate the theory of control ( sliding mode control and Neuro-fuzzy-sliding mode control with disturbance rejection law model ). Section 4 a simulation results to confirm the validity of the proposed technique of control. Finally we conclude by a conclusion in section 5 .

## 2 Electric Vehicle Traction System

This section is dedicated to the dynamics of the electric vehicle and the different components on board the vehicle and their equations

### 2.1 Vehicle dynamics

Consider a vehicle of mass, $m_v$ proceeding at a velocity v , up a slope of angle $\alpha$ , as in Fig.1. The force propelling the vehicle forward $F_T$ , the tractive effort, has to overcome the following forces (rolling resistance, aerodynamic resistance, gravitational force) [25]:

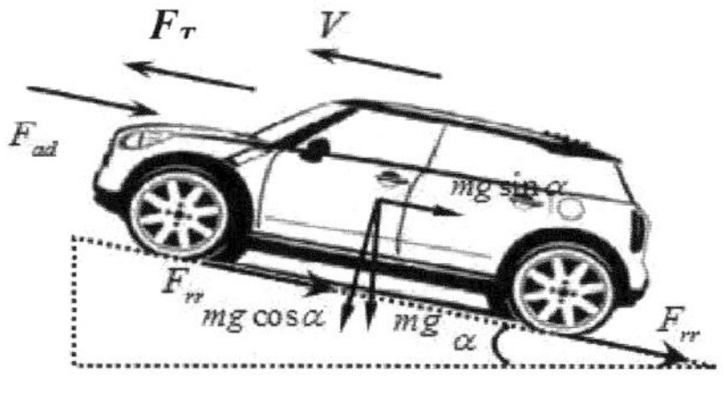


**Fig.1** . Block diagram of the forces acting on the vehicle.

$$F_{rr} = C_{rr}.m_v.g.\cos(\alpha) \tag{1}$$

$$F_{ad} = \frac{1}{2}.\rho.A_f.C_d.(v - v_0)^2 \tag{2}$$

$$F_g = \pm m_v . g . \sin(\alpha) \quad (3)$$

$$m_v \frac{d}{dt} V(t) = F_T(t) - (F_{ad}(t) + F_{rr}(t) + F_g(t)) \quad (4)$$

$$F_{acc} = m_v . \frac{d}{dt} . v(t) \quad (5)$$

Where $C_{rr}$ is the coefficient of rolling resistance, $\rho$ density of the air, $v_0$ the wind speed, $A_f$ the frontal surface of the vehicle, coefficient of penetration into the air which takes values $C_d$ . and the force due to the acceleration $F_{acc}$ ensures the dynamic behavior desired by the driver.

### 2.2 Mathematical model of PMSM

In this work we use the three phases induction motor type PMSM in the stationary (d – q) reference frame, the mathematics mode of permanent-magnet synchronous motor is shown as below [20].

$$\begin{cases} \dot{i}_d = -\frac{R_s}{L_d} i_d + \frac{L_q}{L_d} p\omega_r i_q + \frac{1}{L_d} u_d \\ \dot{i}_q = -\frac{R_s}{L_q} i_q - \frac{L_d}{L_q} p\omega_r i_d - \frac{p\psi_f}{L_q} \omega_r + \frac{1}{L_q} u_q \\ \dot{\omega}_r = \frac{p(L_d - L_q) i_d + p\psi_f}{J} i_q - \frac{1}{J} C_r - \frac{f}{J} \omega_r \end{cases} \quad (6)$$

With $R_s$ :Stator resistance, $L_d$ , $L_q$ d and q axis stator inductances $\psi_f$ :Permanent-magnet flux linkage, $i_d$ , $i_q$ : Stator currents, $u_q$ : Stator voltages, $\omega_r$ : Speed mechanic, $J$ : Moment of inertia, $f$ : Coefficient of viscous friction, $p$ : Number of pole pairs, $C_r$ : Load torque.

### 2.3 Modelling of The Electronic Differential

The proposed structure design propulsion system assures that over a straight trajectory of the wheels ability to roll exactly at the same speed by allowing one to develop an electronic differential. whereas in a curve trajectory; the difference between these two wheel's speed assures the vehicles trajectory over the curve. The fig.2 represents the vehicle structure describing a curve, where $L$ represents the wheelbase, $\delta$ the steering angle, $d$ the distance between the wheels of the same axle and $\omega_{r\_L}$ and $\omega_{r\_R}$ the angular speeds of the left and right wheel drives, respectively.
In a curve the linear speed of each wheel drive is expressed as a function of the vehicle speed and the radius of curve:

$$v_L = \omega_v \left( R + \frac{d}{2} \right) \quad (7)$$

$$v_R = \omega_v \left( R - \frac{d}{2} \right) \quad (8)$$

The curve radius is related to the wheelbase and steering angle:

$$R = \frac{L}{\tan(\delta)} \quad (9)$$

Substituting (9) into Equations (7) and (8), we obtain the angular speed in each wheel drive:

$$\begin{cases} \omega_{rL} = \dfrac{L + 0.5\, d \tan(\delta)}{L} \omega_v \\ \omega_{rR} = \dfrac{L - 0.5\, d \tan(\delta)}{L} \omega_v \end{cases} \quad (10)$$

The difference between the angular speeds of the wheel drives is expressed by the relation

$$\Delta\omega = \omega_{rR} - \omega_{rL} = \frac{d \tan(\delta)}{L} \omega_v \quad (11)$$

Numeric sign of the steering angle signal indicates the curve direction:

$\delta \succ 0 \Rightarrow$ Turn right

$\delta = 0 \Rightarrow$ Straight ahead

$\delta \prec 0 \Rightarrow$ Turn left.

When the vehicle begins a curve, the driver imposes a steering angle to the wheels. The electronic differential however acts immediately on the two motors reducing the speed of the inner wheel ,increases the speed of the outer wheel. The driving wheel angular speeds are:

$$\begin{cases} \omega_{r\_L\_ref} = \omega_v + \dfrac{\Delta\omega}{2} \\ \omega_{r\_R\_ref} = \omega_v - \dfrac{\Delta\omega}{2} \end{cases} \quad (12)$$

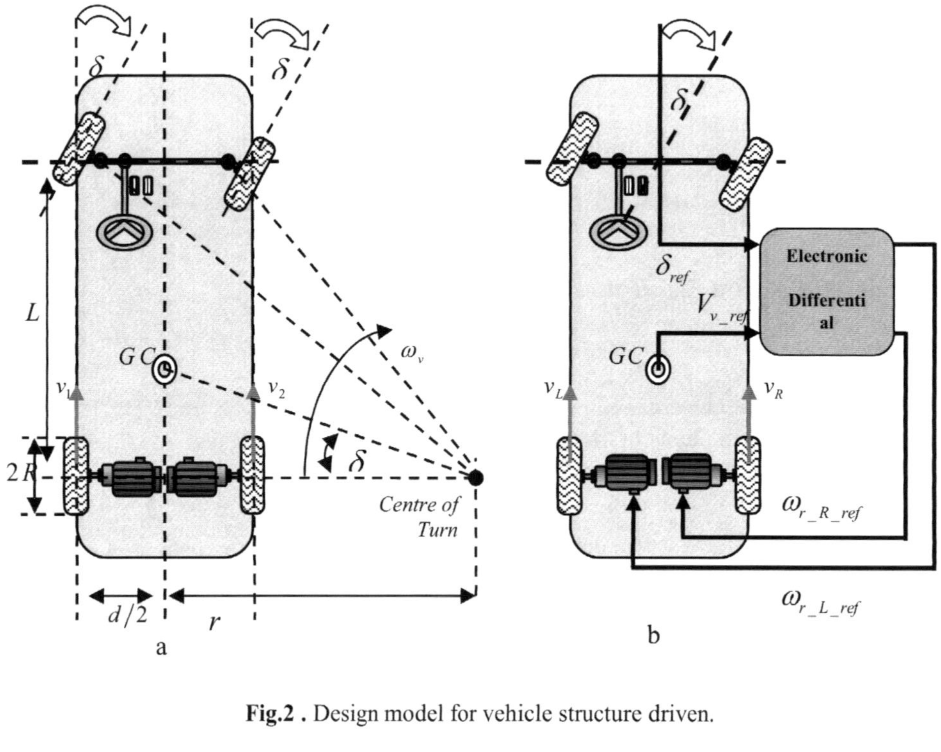


**Fig.2 .** Design model for vehicle structure driven.

## 3. Theory of the Control Used in the System

Fig.3 shows scheme of the proposed control method of the electric traction system using the principle of the cascade control method, the structure comprises a speed control loop which generates the current reference $i_{qref}$ which imposes the control $v_{qref}$, the control $v_{dref}$ is imposed by the current regulation $i_{dref}$

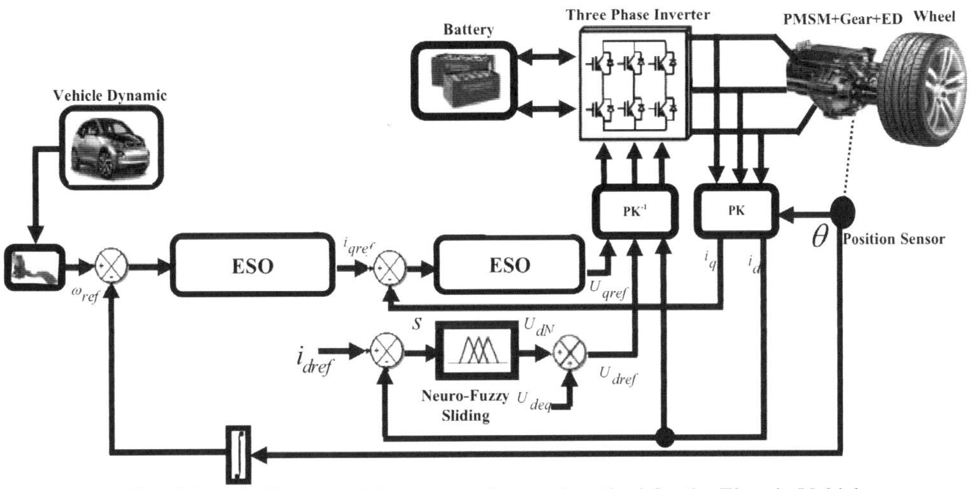


**Fig. 3.** Block diagram of the proposed control method for the Electric Vehicle

### 3.1 Adaptive Neuro-Fuzzy Sliding Mode Controller

The phenomenon of sliding modes could appear in a dynamic system restrained by ordinary differential equations with discontinuous right-hand sides. This phenomenon may happen if the control of the system state switches at high frequency.

However, like anything else the SMC has its own disadvantages, that is to say chattering phenomenon, which initiated from the interaction between parasitic dynamics and high frequency switching control. As stated in [15] and [23] several control methods were suggested in literature to prevent this problem.

We take the general equation suggested by J.J.Slotine to determine the sliding surface :

$$s(x,t) = \left( \frac{\partial}{\partial t} + \lambda \right)^{(n-1)} .e \tag{13}$$

$s$ : the sliding surface

$e$ : Error of the quantity to be controlled,

$\lambda$: Vector of slopes of the $s$ ,

$n$ : Relative degree, equal to the number of times it derives the output for the command to appear.

Sliding mode control of the electric traction system applies the principle of the cascade control method, the structure includes a speed control loop which generates the current reference $i_{qref}$ which imposes the control $v_{qref}$, the control $v_{dref}$ is imposed by the current regulation $i_{dref}$ .

During the sliding mode on direct current loop we have $s^{\bullet}(i_d) = 0, U_{dN} = 0$

$$U_{deq} = (i^{\bullet}_{dref} + \frac{R_s}{L_d} i_d - \frac{L_q}{L_d} p\omega_r i_q) L_d \tag{14}$$

$$U_{dN} = K_d sign(s(i_d)) \tag{15}$$

Where : $k_d$ is positive gain for the direct current regulator.

The proposed Neuro-fuzzy-sliding mode control (NFSMC) replaces the sign function of the direct current surface. Among many, the ANFIS (Adaptive Neuro Fuzzy Inference System) is typically architectured as demonstrated in Fig.4 where a circle indicates a fixed node, however a square indicates an adaptive node. To simplify things, we consider two inputs x, y and one output z among plenty FIS models [23], the Sugeno fuzzy model is the most widely applied. First due to its high interpretability, second computational efficiency, and built-in optimal and adaptive techniques as founded in [21].

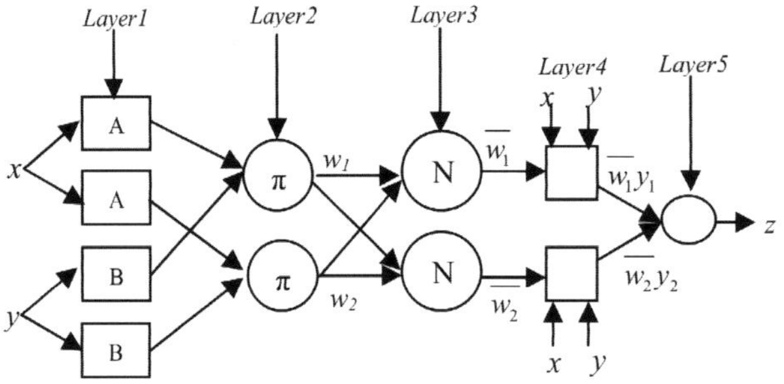


**Fig. 4.** Architecture of ANFIS

based on direct current error ( $e$ ) defined as [23]:

$$e = i_{dref} - i_d \tag{16}$$

Where $i_{dref}$ and $i_d$ are the references and the actual direct current, respectively.

In this study first order Sugeno type fuzzy inference was used for ANFIS and the typical fuzzy rule is:

if $e$ is Ai and $de$ is Bi then y=f( $e$ ) .

Where A and B are fuzzy sets in the antecedent and y=f( $e$ ) is a crisp function in the consequent.

The significances of ANFIS structure are:

Layer 1: Each adaptive node in this layer generates the membership grades for the input vectors Ai, i=1,…,3. In this paper, the node function is a triangular membership function:

$$O_i^1 = \mu_{Ai}(e) = \begin{cases} 0 & e \le a_i \\ \dfrac{e-a_i}{b_i - a_i} & a_i \le e \le b_i \\ \dfrac{c - e}{c - b_i} & b_i \le e \le c_i \\ 0 & c_i \le e \end{cases} \tag{17}$$

Layer 2: The total number of rule is 9 in this layer. Each node output represents the activation level of a rule:

$$O_i^1 = w_i = \min((\mu_{Ai}(e)) \quad i = 1....3 \tag{18}$$

layer 3:Fixed node i in this layer calculate the ratio of the i-th rule's activation level to the total of all activation level:

$$O_i^3 = \overline{w_i} = \frac{w}{\sum_{j=1}^{n} w_j} \tag{19}$$

Layer 4: Adaptive node i in this layer calculate the contribution of i-th rule towards the overall output, with the following node function:

$$O_i^4 = \overline{w_i} y_i = \overline{w_i}(p_i e + r_i) \tag{20}$$

Layer 5: The single fixed node in this layer computes the overall output as the summation of contribution from each rule:

$$O_i^5 = \sum_{i=1}^{2} \overline{w_i} y_i = \frac{w_1 y_1 + w_2 y_2}{w_1 + w_2} \tag{21}$$

The parameters to be trained are, ai, bi, and ci of the premise parameters and pi and ri of the consequent parameters. Training algorithm requires a training set defined between inputs and output [23].

### 3.2 Extended State Observer Controller

Parameters variation, back electromotive forces, torque ripples, friction forces, unmodeled dynamics, external disturbance and load disturbances. All these different kinds of disturbances appear in real applications of electric vehicle based on PMSM motors. If the controller does not have adequate ability to these disturbances , they may demean the performance of the electric vehicle system.

Since the control structure that has been employed here is of cascade control loops for the speed regulation problem, the control scheme includes a speed loop and two current loops. For the d-axis current loop, Adaptif Neuro Fuzzy Sliding control algorithm is employed, for the speed loop and q-axis current loop, in order to estimate the different disturbances in each loop correspondingly, two ESO's are being introduced.

The lumped disturbances of the system which are composed of internal dynamics and external disturbances are considered as new state of the system by Extended State Observer (ESO) [18], where both states and lumped disturbances $a\,(t)$ can be estimated by this observer.

The construction of ESO will be shown below :

$$x^{\bullet} = f(x,t) + d(t) + bu(t) \tag{22}$$

where $f(x,t)$ is the unknown nonlinear function, $d(t)$ is the external disturbance, $u$ is the control signal, $x$ is the output of system, $b$ is the control gain. Considering the lumped disturbance $a(t) = f(x,t) + d(t)$ as an extended state, then Eq. (29) can be expressed as:

$$x^{\bullet} = a(t) + bu(t) \tag{23}$$

Define $x_1 = x$, $x_2 = a(t)$ . The system (30) can be rearranged as:

$$\begin{cases} x_1^{\bullet} = x_2 + bu(t) \\ x_2^{\bullet} = c(t) \end{cases} \tag{24}$$

where $c(t) = a^{\bullet}(t)$. And the ESO can be constructed as follows [26]:

$$\begin{cases} z_1^{\bullet} = z_2 + 2P(z_1 - \omega_r) + bu(t) \\ z_2^{\bullet} = -P^2(z_1 - \omega_r) \end{cases} \tag{25}$$

where $-p(p > 0)$ is the desired double pole of ESO. $z_1$ is the estimate of x1 and $z_2$ is the estimate of $a(t)$. Here, according to the analysis in [26], $z_1(t)$ is an estimate of the output $x(t)$ while $z_2(t)$ is an estimate of the lumped disturbance $a(t)$, i.e., $z_1(t) \to y(t)$ and $z_2(t) \to a(t)$.

A composite control law is designed as follows:

$$u = u_0 - \frac{z_2}{b},\ u_0 = k(x_{ref} - z_1) \tag{26}$$

where $x_{ref}$ is the reference input of $x$.

### *3.2.1 ESO Speed controller*

ESO-based controller for the speed loop is designed as follows:
The torque equation of PMSM system is:

$$\omega_r^{\bullet} = \frac{3p\psi_f}{2J} i_q - \frac{f}{J}\omega_r - \frac{C_r}{J} = bi_{qref} + a(t) \tag{27}$$

Where $b = \frac{3p\psi_f}{2J}$ is the control gain, $a(t) = \frac{3p\psi_f}{2J}(i_q - i_{qref}) - \frac{f}{J}\omega_r - \frac{C_r}{J}$ is the lumped disturbances consisting of the friction, the external load disturbances, and the tracking error of current-loop of $i_q$.

Then an ESO can be designed as:

$$\begin{cases} z_1^{\bullet} = z_2 + 2P(z_1 - \omega_r) + bu(t) \\ z_2^{\bullet} = -P^2(z_1 - \omega_r) \end{cases} \tag{28}$$

where $z_1$ is the estimate of angular velocity $x$, $z_2$ is the estimate of the lumped disturbances.

A composite control law of speed loop can be designed as:

$$i_{qref} = u = u_0 - \frac{z_2}{b},\ u_0 = k(\omega_{rref} - \omega_r) \tag{29}$$

where $k$ is the proportional gain, $\omega_{rref}$ is the reference speed, $i_q$ is the output of speed loop controller.

The block diagram of the ESO-based speed controller is shown in Fig.5. Note that the ''generalized System'' represents the two current loops which include PMSM and other components the same as that of Fig.3.

### *3.2.2 ESO Iq Current controller*

The ESO of q-axis current loop can be designed as follows:

$$i_q^{\bullet} = -p\omega_r i_d - \frac{R_s}{L} i_q - \frac{p\psi_f}{L}\omega_r + \frac{u_q}{L} \tag{30}$$

$$\dot{i}_q = a_q(t) + \frac{u_q}{L} \tag{31}$$

the lumped disturbance of q-axis current loop are $a_q = -p\omega_r i_d - \frac{R_s}{L} i_q - \frac{p\psi_f}{L}\omega_r$ consisting of the coupling between rotor speed and d-axis currents, the dynamics of q-axis current, the back electromotive force .

$$\begin{cases} \dot{z}_{11} = z_{12} + 2P_q(z_{11} - i_q) + \frac{u_q}{L} \\ \dot{z}_{12} = -P_q^2(z_{11} - i_q) \end{cases} \tag{32}$$

Where $-P_q(P_q \succ 0)$ is the desired double pole , $z_{11}$ is the estimated of $i_q$ ,and $z_{12}$ is the estimated of $a_q(t)$.

A hybrid control law of q-axis current loop can be designed as :

$$u_q = u_{q0} - Lz_{12},\ u_{q0} = k_q(i_{qref} - i_q) \tag{33}$$

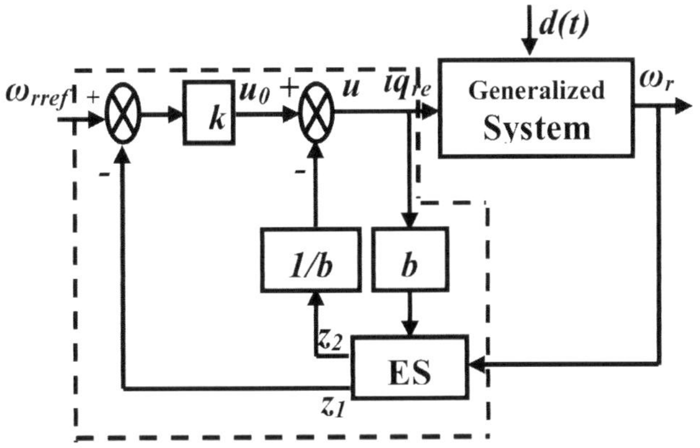


**Fig.5.** The bloc diagram of ESO-based speed controller

## 4. Simulation Results

It is important to note that the simulation requested by the driver, which demoes the wheel's speed while steering for 0 < t < 481 run under a Japanese urban driving cycle 11 mode 4.

With the vehicle approaching a right turn at 62 s (positive steering angle) and a left turn at 180 s (negative steering angle) .It is evident that the electric vehicle adequately functioning according to the complicated series of accelerations, decelerations, and frequent stops.

Concerning the figure 6 and 7 , the vehicle is turning right at t= 62 s,the speed left motor increases and right speed motor decreases, but the torque of the right wheel is increasing negatively which is basically a braking mode. As we can observe that the torque is returning to its initial value at the moment the speed of the right wheel is stabilized , where this working phase can be exploited for energy recuperation.

As we can view, the driver uses an inverse steering angle on the front of wheels at t= 67 s in order to exit the curved section of the road. The electronic differential subsequently acts to equalize the speed of the drive motors; the speed difference of the driving wheels is shown in figure 6. Unlike that on the right, the electronic differential is acting at once on both electric motors by lowering the speed of the left wheel, thus the vehicle approaching a left turn at t=180s . This speed is perceived in figure 6 .

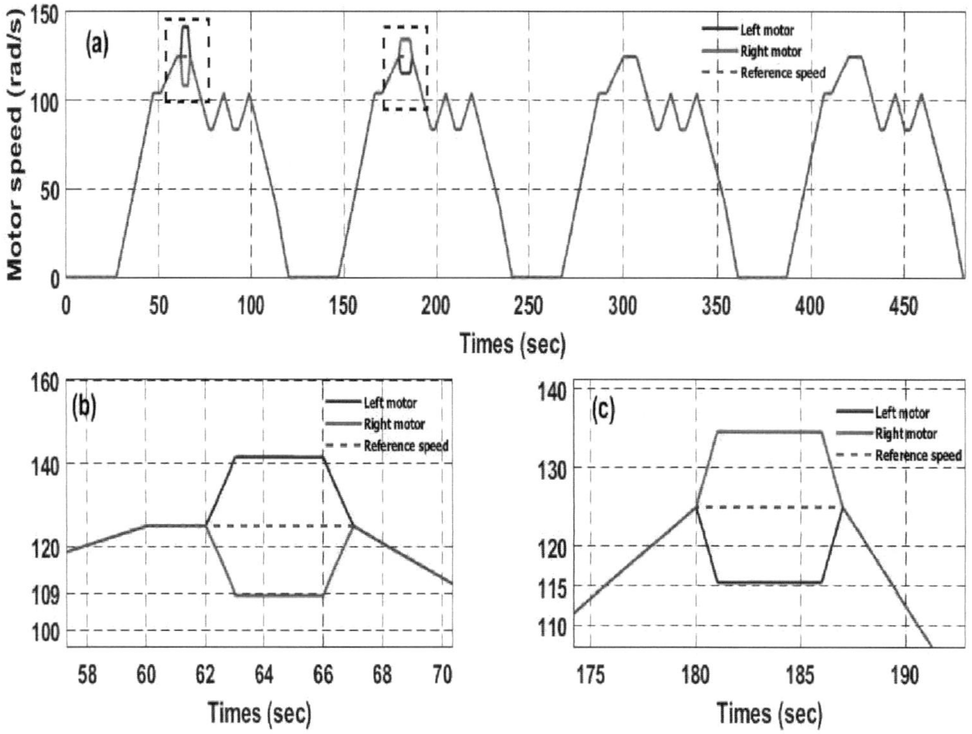


**Fig. 6.** (a): Speed motor of the vehicle, (b): zoom around 65 s ,(c):zoom around 186s

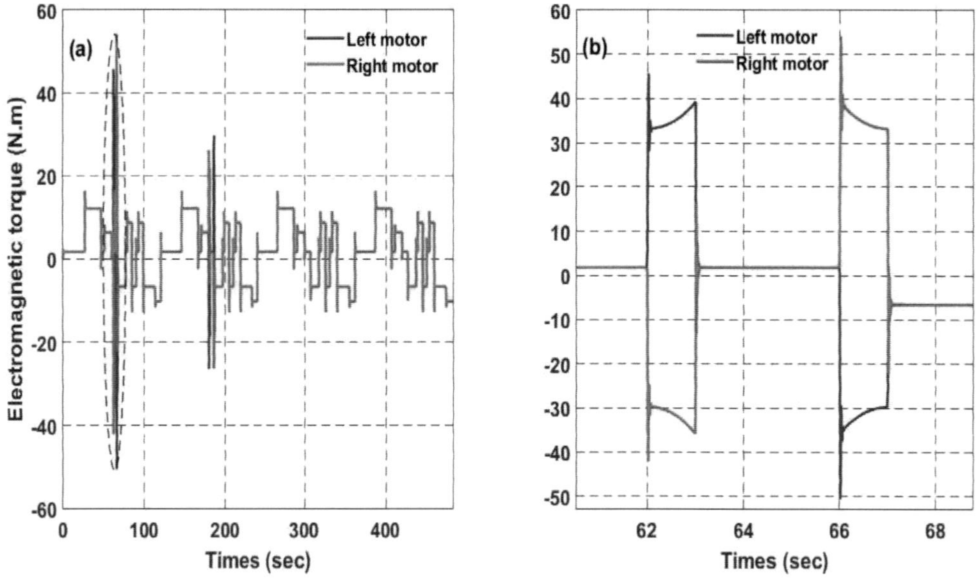


**Fig. 7.** (a):Variation of electromagnetic torque of the right and left motor in different phases,(b): zoom around 65 s

## 5. CONCLUSION

In this paper, an Electronic Differential And Neuro-Fuzzy Sliding Mode Control With Extended State Observer For An Electric Vehicle System has been designed. The EV stability can be hard to manage particularly while cornering or even under slippery road condition. For that purpose, the recommended Electronic Differential (ED) has been developed. the problem of chattering phenomena in the PMSM electric motor had been eliminated by employing of tow types of controllers, neuro-fuzzy sliding mode controller on the direct current loop, ESO controller on the speed and quadratic current loops; taking into account the dynamic of the vehicle.T he simulations results indicate on the one hand that the traction system using NFSMC with ESO exhibits better dynamic performance, static performance, and robustness than the system with the traditional SMO, NFSMC with ESO has smaller chattering, higher accuracy, and no phase delay; and on other hands the electronic differential controls the driving wheels speeds with high accuracy either in flat roads or curved ones. In future studies, the proposed design technique will be extended to the improved ESO control strategy of current and position controls.

## Referencesm

1. Michele D G , Elena P, Giorgio M , Energy Feb.2015;82:294-311.
2. Kandasamy N K , King J T , Energy Jun.2016;111:190-196.
3. A. Roque, J.Esteves, J.Maia, P.Verdelho, Proceeding of Electrimacs **99** September (1999)
4. M. Martin , *Info Tech n°13,* June (2004)
5. K. Houacine, R. Mellah and S. Guermah, International Journal of Electric and Hybrid Vehicles, vol. 7, pp. 189–207, 2015.
6. S. Morimoto et al., IEEE Transactions on Industrial Electronics, vol.61 , no.10, pp. 5749-5756, Oct. 2014.
7. A. Sant and K. Rajagopal , IEEE Transactions on Magnetics, vol.45, no.10, , pp. 4672-4675, Oct. 2009.
8. Golea N, Golea A, Kadjoudy M. IEEE international conference on industrial technology. Mumbai, India; 2006.p p. 1533–7.
9. Hsien TL, Sun YY, Tai MC, IEE Proc – Electr Power Appl 1997;144(3):173–81.
10. Wai RJ. , IEEE Trans Indust Electron 2001;8(5):926–44.
11. Wang GJ, Fong CT, Chang KJ. , IEEE Trans Indust Electron 2001;48(2):408–15.
12. Kung YS, Tsai MH., IEEE Trans Power Electron 2007;22(6):2476–86.
13. Max Reitz , Xin Wang, Patrick Gu, 978-1-5090-0403-4/16,2016,IEEE.
14. Chung, S.C.Y., Lin, C.L., IEEE Trans. Automatic Control 44, 563–568 (1999).
15. N.K. Yadav and R. K. Singh , International Journal of Artificial Intelligence & Applications, Vol. **4**, July (2013).
16. L. Jian, Z. Wang , L. Jian, Z. Wang , Microcomputer Information, Vol. **23**, n°. **9**, pp. 22-24, (2007)
17. H. Bai, R. Qi, Transactions of China Electrotechnical Society, Vol. **23**, n°. **8**, pp. 36-40, (2008)
18. J. Q. Han, Control Decision, vol. **10**, pp. 85–88, 1995, (in Chinese).
19. C.C. Chan, Y.S. Wong, IEEE Power and Energy Magazine, Vol. **2**, No. **6**, Nov/Dec. (2004), pp. 24 – 33.
20. R.F. Fung, C.L. Chiang, G.C. Wu, Appl. Math Model. **34** (2010) 2323–2335

21. J.S.R. Jang, IEEE Trans Syst. Man. Cybernet, 1993; **23(3)**, pp. 665-8
22. D. Foito , A. Roque, J.Esteves, J.Maia, Proceeding of EVS17, Montreal , October (2000)
23. A.Tahour, A.Hamza , A.Aissaoui, Advance in Electrical and Computer Engineering, Vol,8.15,Number **1(29)**,(2008)
24. A.Tahour , H.Abid, A.Aissaoui, serbian journal of electrical engineering Vol. **4**, No. 1, June (2007), 23–34
25. Sadeghi, S. & Mirsalim, M, International peer-reviewed scientific journal of Applied sciences, Vol.7, No.1, (2010), pp. 51-71, ISBN 1785-8860.
26. Miklosovic R, Gao ZQ, In: 39th IAS annual meeting on industry application conference; 2004. p. 1495–502.

# Assessment of encapsulation materials for solar cells

*Vladimír* Šály, *Vladimír* Ďurman, *Michal* Váry, *Milan* Perný and *František* Janíček

Slovak University of Technology, Faculty of Electrical Engineering and Information Technology
Ilkovičova 3, SK-81219 Bratislava, Slovak Republic

**Abstract.** Interfacial processes were studied in various insulation foils intended for encapsulation of photovoltaic cells. The analysis was based on the dielectric measurements in a broad region of temperatures and frequencies. The measurements showed that the observed processes are connected with the electrode polarization. The electrode polarization gives rise to the space charge formation and enhancement of electric field near the electrodes. Calculation of the electric field is important for praxis as it allows assessing the risk of electrical breakdown. In our work we use the parameters obtained from the dielectric measurements for calculation of electric field distribution in encapsulating materials. It was found that electric field increases more than 100-times comparing with the mean value.

## 1 Introduction

Compact photovoltaic module is a layered structure consisting of interconnected and encapsulated solar cells embedded between substrate and superstrate (back and front sheets), glass or plastic sheets, in order to create compact and sealed unit protected against environmental influences. Solar cells strings are encapsulated between two sheets of encapsulant foil. Encapsulant is a polymeric material which also provides adhesion between substrate sheet and superstrate glass and strings of solar cells among them [1]. Photovoltaic module is source of electricity and from the electrical point of view encapsulant insulates electrically active solar cells and interconnecting wires on one side and also protects against direct electric contact in order to avoid electric hazard. Insulation properties and dielectric properties are among the important indicators tested at the beginning and also during life-time of photovoltaic module [2]. Dielectric and electric tests are common to ensure the efficient operation and safety over time [3, 4]. The encapsulant as an electric insulator has to withstand electric stress which when steadily increased, finally overcomes electric strength. The processes leading to electric breakdown are either quite rapid – short term or longer term caused by eg partial discharge activity or electrical treeing [5]. Electrical faults occurring in PV arrays as a result of electric breakdown are more hazardous that demand immediate attention [6]. In the case of many solar cells connected into series, the encapsulant is electrically stressed mainly at sharp edges of the structure. Another source of enhanced electric field is accumulation of space charge at the dielectric – electrode interfaces. This phenomenon is connected with the electrode polarization [7]. In the process of polarization the free cations in dielectric are accumulated near the negative electrode, creating a positive charge. A similar process runs near the positive electrode. The steady state of accumulation is given by the balance of concentration gradient and the Coulombic (electric) force [8, 9]. The cations in dielectric and the negative charge at the electrode create so-called double layer. The layer is characterized by the Debye length:

$$L_D = \frac{1}{q}\left(\frac{\varepsilon_0 \varepsilon_R kT}{n_0}\right)^{1/2}, \tag{1}$$

where $q$ is the ionic charge, $\varepsilon_R$ is the relative permittivity of material in absence of free charges, $n_0$ is the free ion concentration, $k$ is Boltzmann's constant and $T$ is temperature. In the case of non-polar material the value of $\varepsilon_R$ equals to $\varepsilon_\infty$ (the optical permittivity). The whole process matches to the Debye relaxation function with the static permittivity of

$$\varepsilon_s = \frac{L}{2L_D}\varepsilon_\infty = \frac{L\varepsilon_\infty}{\frac{2}{q}\left(\frac{\varepsilon_0 \varepsilon_\infty kT}{n_0}\right)^{1/2}}. \tag{2}$$

For relaxation time $\tau_0$ the next equation is valid:

$$\tau_0 = \frac{L}{2L_D}\frac{\varepsilon_0 \varepsilon_\infty}{\sigma}. \tag{3}$$

In the previous equations we used label $L$ for the electrode distance and $\sigma$ for the material conductivity.

## 2 Encapsulants

The matter of investigation were three different materials commonly used in PV module production as encapsulants of solar cells. They were namely ethylene vinyl acetate (EVA), polyvinyl butyral (PVB) and thermoplastic polyurethane (TPU). Their composition depends on the producer and the final product should be stabilized and resistant against environmental influences with appropriate electric properties. EVA is most frequently used. It is copolymer of ethylene and vinyl acetate. The properties of EVA are improved with additives. [10]. Some details of composition and the function of respective additives is discussed in [11]. PVB is mostly composed of 76 % vinyl butyral, 22 % vinyl alcohol and 2 % vinyl acetate [12] originally more intended for safety glass. TPU represents relatively new material exploited as photovoltaic encapsulant composed of softeners, solid segments and filling additives. The properties and advantages in comparison to the other encapsulants are more precisely discussed e.g. in [13].

## 3 Experiment

In our work we examined three types of encapsulating materials which are regularly used for solar cells. The dielectric measurements were carried out with emphasis on the region of very low frequencies, where the electrode polarization can be identified. The measured data were treated in the form of frequency dependence of the absolute value of impedance and its phase. The experiments were made by an impedance analyzer constructed at our department particularly for the range of very low frequencies. It is based on measuring the voltage on the divider consisting of the impedance normal and the measured object. The real and the imaginary part of the measured impedance are calculated by the Fourier transform of the mentioned voltages. This transform can also reveal a possible non-linear behavior of the measured structure. The actual frequency range of this equipment is from 0.001 Hz to 1 kHz. The voltage on the specimen during measurements was 5 V. The specimens were investigated in the temperature range from about 30 to 80 °C. During the measurements the temperature was stabilized in the thermostat TETTEX, type 2965. We used the stainless steel three-electrode system with electrode diameter of 5 cm. The thicknesses of the foil specimens were: 0.50 mm (EVA), 0.36 mm (PVB) and 0.47 mm (TPU).

The measured values of impedance are compared by the means of the phase plots in Figs. 1 – 3. The phase obviously changes from zero at low frequencies to -π/2 at higher frequencies. In the case of electrode polarization, the phase approaches to -π/2 on both sides of the frequency region. It gains a very flat maximum with the value near to zero. A dipolar polarization (if exists) causes a small maximum of phase in the medium or high frequency range. From the results of our measurements we can state, that no polarization can be identified in the EVA specimen. The phase in this material shows only a conduction behavior (Fig. 1). The electrode polarization can be seen in the PVB specimen with expressive maximum of phase at 0.1 Hz (Fig. 2). Electrode and dipolar polarizations are present in TPU foil where the small maximum of dipolar polarization is overlapped by the electrode process (Fig. 3).

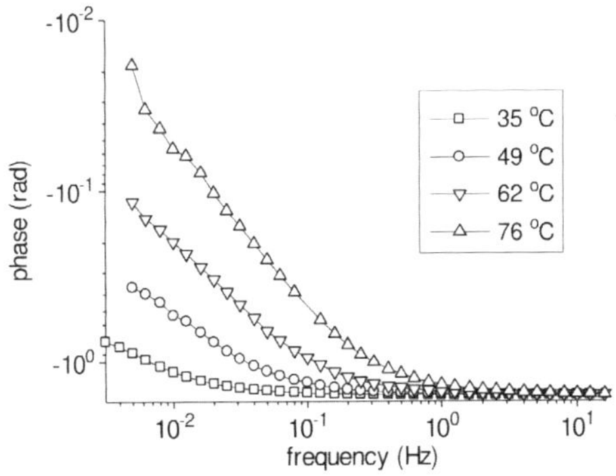


**Fig. 1.** Phase plot for EVA.

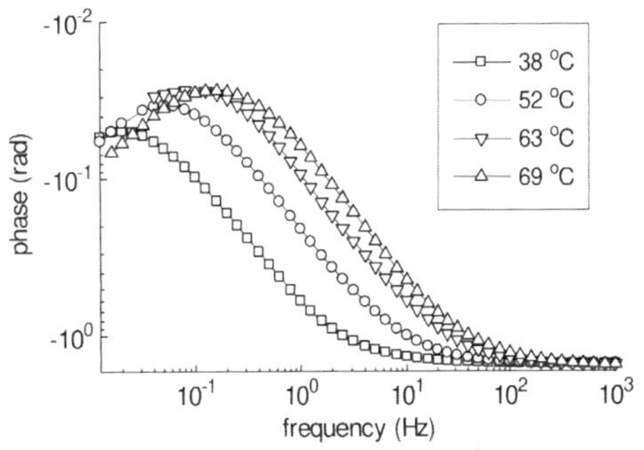


**Fig. 2.** Phase plot for PVB.

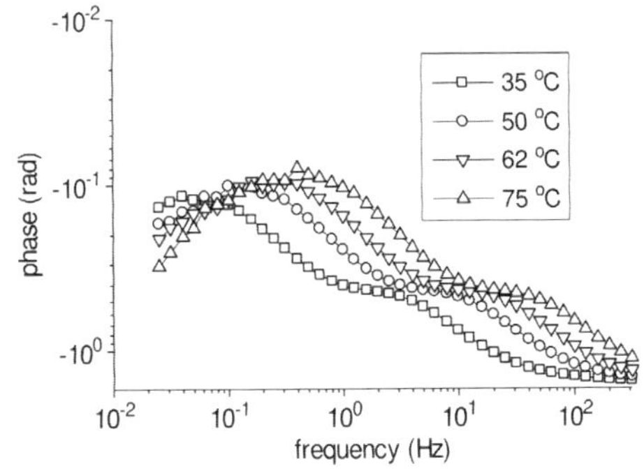


**Fig. 3.** Phase plot for TPU.

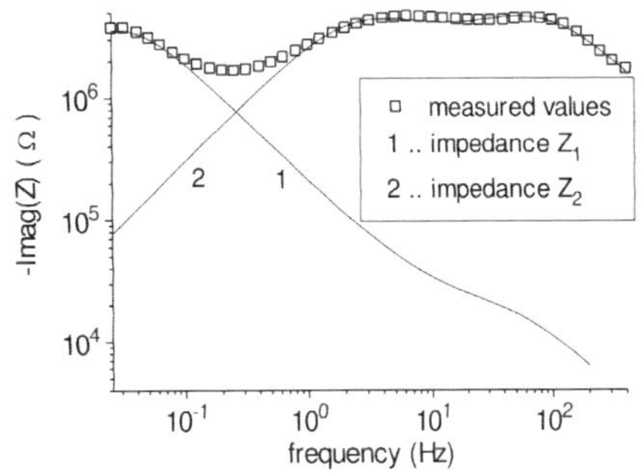


**Fig. 4.** Fitting functions - TPU (62 °C).

## 4 Results and discussion

At modeling the dielectric processes in PVB and TPU encapsulating materials we come out from the assumption, that the thin Debye layer is arranged in parallel with the dielectric in the bulk of specimen. Both layers have the same dielectric parameters, but they differ in their conductivities. We are dealing with two impedances which are described with help of the well-known Cole-Cole formula [14]:

$$Z_I = \frac{1}{j\omega C_{0I}\left(\varepsilon_\infty + \dfrac{\varepsilon_s - \varepsilon_\infty}{1+(j\omega\tau_0)^{1-\alpha}} - j\dfrac{\sigma_I}{\varepsilon_0\omega}\right)}. \tag{4}$$

The index $I$ in the above formula acquires alternatively value 1 (for Debye layer) or 2 (for bulk layer). Here we used the next labeling: **j** is the imaginary unit, $\omega$ is the angular frequency, $\varepsilon_\infty$ is the optical permittivity of the dielectric, $\varepsilon_0$ is the vacuum permittivity, $C_0$ is the geometric or vacuum capacitance, $\varepsilon_s$ is the static permittivity, $\tau_0$ is the relaxation time, $\alpha$ is the shape parameter and $\sigma$ is the conductivity of dielectric. The unknown material parameters were calculated by fitting the measured values of the complex impedance $Z$ with the sum of impedances $Z_1$ and $Z_2$. This was done by means of the Nelder-Mead type simplex search method. An example of fitting the imaginary part of impedance for TPU foil at 62 °C is in Fig. 4. The complete calculated parameters are in Tab.1 (for PVB) and Tab. 2 (for TPU). The values are listed as functions of temperature $t$.

**Table 1.** Dielectric parameters of PVB encapsulating foil.

| t (°C) | $\varepsilon_\infty$ | $\varepsilon_s$ | $\tau_0$ (s) | $\sigma_1$ ($10^{-12}$ Sm$^{-1}$) | $\sigma_2$ ($10^{-9}$ Sm$^{-1}$) | $h_1$ ($10^{-7}$ m) | $h_2$ ($10^{-4}$ m) | $E_1/E_2$ |
|---|---|---|---|---|---|---|---|---|
| 39 | 5.76 | 8.76 | 0.176 | 4.43 | 0.51 | 4.57 | 3.60 | 54 |
| 45 | 5.76 | 8.67 | 0.118 | 6.14 | 0.89 | 1.81 | 3.60 | 144 |
| 52 | 5.75 | 8.56 | 0.075 | 6.34 | 1.65 | 1.06 | 3.60 | 260 |
| 58 | 5.75 | 8.47 | 0.052 | 7.46 | 2.75 | 0.78 | 3.60 | 369 |
| 63 | 5.74 | 8.40 | 0.039 | 8.91 | 4.16 | 0.64 | 3.60 | 466 |
| 69 | 5.74 | 8.32 | 0.027 | 1.13 | 6.70 | 0.52 | 3.60 | 590 |
| 75 | 5.73 | 8.24 | 0.020 | 1.47 | 10.60 | 0.44 | 3.60 | 722 |

**Table 2.** Dielectric parameters of TPU encapsulating foil.

| t (°C) | $\varepsilon_\infty$ | $\varepsilon_s$ | $\tau_0$ (s) | $\alpha$ | $\sigma_1$ ($10^{-12}$ Sm$^{-1}$) | $\sigma_2$ ($10^{-9}$ Sm$^{-1}$) | $h_1$ ($10^{-7}$ m) | $h_2$ ($10^{-4}$ m) | $E_1/E_2$ |
|---|---|---|---|---|---|---|---|---|---|
| 35 | 6.61 | 37.8 | 0.135 | 0.07 | 2.17 | 1.74 | 18.0 | 4.69 | 80 |
| 42 | 6.48 | 37.7 | 0.088 | 0.07 | 2.70 | 2.76 | 14.1 | 4.69 | 102 |
| 50 | 6.34 | 37.6 | 0.055 | 0.08 | 3.59 | 4.56 | 11.2 | 4.69 | 127 |
| 56 | 6.24 | 37.6 | 0.039 | 0.09 | 4.50 | 6.55 | 9.77 | 4.69 | 145 |
| 62 | 6.14 | 37.5 | 0.028 | 0.09 | 5.67 | 9.27 | 8.64 | 4.69 | 163 |
| 68 | 6.05 | 37.4 | 0.020 | 0.10 | 7.14 | 13.0 | 7.75 | 4.69 | 181 |
| 75 | 5.95 | 37.3 | 0.014 | 0.11 | 9.33 | 18.9 | 6.91 | 4.69 | 202 |

The value of $\alpha$ was negligible for PVB foil, so it is omitted in Tab. 1. Except the mentioned values, in the tables there are calculated thicknesses of Debye layer (assigned as $h_1$) and the bulk layer ($h_2$). The last columns of the tables include the value of electric field ($E$) enhancement in the Debye layer as $E_1/E_2$ which equals to $\sigma_2/\sigma_1$. This equality comes from the general principle of electric current continuity.

The dielectric and electrode processes in the individual encapsulating materials can be assessed according to the value of conductivity and its activation energy. The conductivity obeys the well-known Arrhenius law, as it is seen in Fig. 5. The calculated activation energies are in Tab. 3.

The processes we have observed can be divided into the part of dipolar relaxation and the part of electrode polarization. The dipolar relaxation appears both in PVB and TPU foil. These processes were already known [15, 16]. They are attributed to the retard motion of the polar side groups on the long polymer chain. When comparing the tested encapsulating materials according to their conductivities, apparently, the EVA foils provides the smaller conductivity with higher activation energy. Conductivity of EVA is approximately 1000 times lower than those of PVB and TPU. On the other hand, the relative high conductivities of the PVB and TPU foil results in the electrode polarization which enhances the local electric field. The thickness of the Debye layer is of the order $10^{-7}$ m while the conductivity at the electrode boundary is more than 100 times lower than in the bulk of material. As the boundary conductivity depends less on temperature than the bulk one, the difference against the bulk conductivity increases with temperature. The ratio of electric field in the Debye layer and the bulk layer in the steady state (dc voltage) is inversely proportional to the ratio of the layers conductivities. That is why the enhancement of electric field is more significant at higher temperatures. As it is seen from the last column of Tab.1 and Tab. 2, the enhancement has approximately linear character with slope 18 for PVB and 3 for TPU. The electrode process of course depends on the electrode material. The electric field enhancement in particular cases can differ from our observations which were realized on stainless steel electrodes.

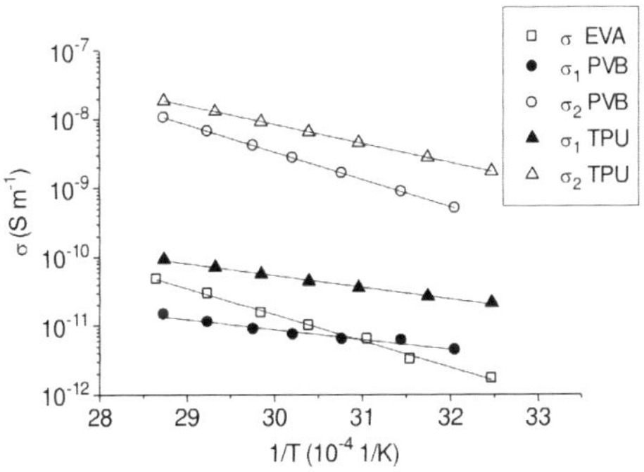


**Fig. 5.** Arrhenius plot of conductivity.

**Table 3.** Activation energies of conductivity.

| Activation energy (eV) | | | | |
|---|---|---|---|---|
| EVA | PVB | | TPU | |
| | $\sigma_1$ | $\sigma_2$ | $\sigma_1$ | $\sigma_2$ |
| 0.76 | 0.29 | 0.79 | 0.34 | 0.55 |

## 5 Conclusion

Three types of insulation materials intended for use as encapsulating foils of solar cells were studied. The investigations were based on the results of impedance spectroscopy. A special importance was attached to the region of very low frequencies. Except of the previously known dielectric relaxation, we detected appearance of electrode polarization in two materials, namely PVB and TPU. By analyzing the impedance data we found the thicknesses of the boundary layers and also their conductivities. These results are important at assessment of electric field distribution during the solar cell operation. The electric field in this configuration is inversely proportional to the ratio of the layers conductivities. At room temperature the field enhancement reaches the value of about 50. At higher temperatures, which are solar cells operational temperatures, increases the field approximately 200 – 600 times. Although the insulating material is stressed only locally, deterioration due to enhanced stress may appear during a long-time operation. In-homogenous electric field distribution represents increased probability to activate the processes leading to deterioration of insulator which can lead at critical level to electric breakdown process. That is why the material must be checked by accelerated aging tests performed at higher electric fields.

The acknowledgements: This work was supported by the Slovak Research and Development Agency (APVV) under the Contract No APVV-15-0326. This contribution is also the result of the project implementation: National Centre for Research and Application of Renewable Energy Sources (ITMS: 26240120016), supported by the Research & Development Operational Programme funded by the ERDF.

## References

1. R. Polanský, M. Pinkerová, M. Bartůňková, P. Prosr, Mechanical behavior and thermal stability of EVA encapsulant material used in photovoltaic module, Journal of Electrical Engineering Vol. **64**, 361 (2013)
2. M. De Oliveira, A. Diniz, M. Viana, V. De Freitas Cunha Linsa, The causes and effects of degradation of encapsulant ethylene vinyl acetate copolymer (EVA) in crystalline silicon photovoltaic modules: A review, Renewable and Sustainable Energy Reviews **81,** 2299 (2018)
3. J. D. Lind, Photovoltaic panel impulse voltage testing, evaluation engineering, Jan. 2011 (https://www.evaluationengineering.com/photovoltaic-panel-impulse-voltage-testing)
4. G. Cibira, PV cells electrical parameters measurement, Journal of Electrical Engineering Vol. **68** 74 (2017)
5. B. Sultan, J. Andersonn, T. Hjertbeg, Electrical breakdown testing of materials intended for use in PV modules, (3rd Atlas/NIST Workshop on Photovoltaics Materials Durability. Gaitherburg, Maryland December 8-9, 2015)
6. D. S. Pillai, N. Rajasekar, A comprehensive review on protection challenges and fault diagnosis in PV systems. Renewable and Sustainable Energy Reviews **91,** 18 (2018)
7. R. Coelho: Sur la relaxation d'une charge d'espace. Revue de Physique Appliquée **18**, 137 (1983)
8. S. Emmert, M. Wolf, R. Gulich, S. Krohns, S. Kastner, P. Lunkenheimer, A. Loidl, Electrode polarization effects in broadband dielectric spectroscopy. European Physical Journal **B83**, 157 (2011)
9. J.R. Klein, S. Zhang, S. Dou, B.H. Jones, R.H. Colby, J. Runt, Modeling electrode polarization in dielectric spectroscopy. Journal Chemical Physics **124**, 144903 (2006)
10. K. Agroui, G. Collins, Characterisation of EVA encapsulant material by thermally stimulated current technique, Solar Energy Materials and Solar Cells **80**, 34 (2003)
11. De Oliveira, M.C.C. et al., The causes and effects of degradation of encapsulant ethylene vinyl acetate copolymer (EVA) in crystalline silicon photovoltaic modules: A review. Renewable and Sustainable Energy Reviews **81,** 2299 (2018)
12. A. K. Dhaliwal, J. N. Hay, The characterization of polyvinyl butyral by thermal analysis, Thermochimica Acta **391**, 245 (2002)
13. G. Stollwerck, H. G. Stevens, H. J. Goerkes, W. Kaufhold, A. Lyding, H. Peerlings, E. U. Reisner, J. Ufermann, M. Vogt, Vacuum-free lamination of solar modules, (Proc. 20th European Photovoltaic Solar Energy Conference, 2002, 2005)

14. M. Wubbenhorst, J. Vanturnhout, Analysis of Complex Dielectric Spectra. I. One-dimensional derivative techniques and three-dimensional modelling, Journal of Non-Crystalline Solids **305**, 40 (2002)
15. A.M. North, J.C. Reid, Dielectric relaxation in a series of heterophase polyether polyurethanes. European Polymer Journal **8**, 1129 (1972)
16. B.L. Funt, Dielectric dispersion in solid polyvinyl butyral. Canadian Journal of Chemistry **30**, 84 (1952)

# Comparative studies of Cu-Cl Thermochemical Water Decomposition Cyles for Hydrogen Production

*Funmilayo Osuolale*[1,*], *Oladipupo Ogunleye*[1], *Mary Fakunle*[1], *Abdulfataah Busari*[1], *and Yetunde Abolanle*[1]
[1]Department of Chemical Engineering, Ladoke Akintola University of Technology, Ogbomoso, Nigeria.

**Abstract.** This research focuses on thermodynamic analysis of the copper chlorine cycles. The cycles were simulated using Aspen Plus software. All thermodynamic data for all the chemical species were defined from literature and the reliability of other compounds in the simulation were ascertained. The 5-step Cu–Cl cycle consist of five steps; hydrolysis, decomposition, electrolysis, drying and hydrogen production. The 4-step cycle combines the hydrolysis and the drying stage of the 5-step cycle to eliminate the intermediate production and handling of copper solids. The 3-step cycle has hydrolysis, electrolysis and hydrogen production stages. Exergy and energy analysis of the cycles were conducted. The results of the exergy analysis were 59.64%, 44.74% and 78.21% while that of the energy analysis were 50%, 49% and 35% for the 5-step cycle, 4-step cycle and 3-step cycle respectively. Parametric studies were conducted and possible exergy efficiency improvement of the cycles were found to be between 59.57-59.67%, 44.32-45.67% and 23.50-82.10% for the 5-step, 4-step and 3-step respectively. The results from the parametric analysis of the simulated process could assist ongoing efforts to understand the thermodynamic losses in the cycle, to improve efficiency, increase the economic viability of the process and to facilitate eventual commercialization of the process.

## 1 Introduction

The world faces problems with depleting energy resources resulting from worldwide increasing energy consumption due to increasing population and rising living standards. This is in addition to concerns regarding global climate change accruing from energy related activities. Hence, there have been extensive researches and developments of alternative clean energy sources which can be derived from renewable sources. There are various alternative energy options to fossil fuels, including solar, geothermal, hydropower, wind and nuclear energy. However, Hydrogen is a potential alternative that has been viewed as a promising fuel [1].It has some unique characteristics that makes it an ideal

* Corresponding author: fnosuolale@lautech.edu.ng

energy carrier that can satisfy all energy needs [2]. Hydrogen, though abundant on earth does not exist alone, it is produced from compounds that contains it. There are various ways of generating hydrogen for energy usage. Some of the various processes include; the solar based hydrogen production, electrolysis, biological processes, gasification and the thermochemical water decomposition [3,4] Hydrogen production via thermochemical water decomposition is one of the key potential source of energy.

Thermochemical water splitting with a copper-chlorine (Cu-Cl) cycle is a promising process involving intermediate copper and chlorine compounds with a net input of water and heat and a net output of hydrogen and oxygen [5,6]. These chemical reactions form a closed internal loop that recycles all chemicals on a continuous basis, without emitting any greenhouse gases. Thermochemical water decomposition is an emerging technology for large-scale production of hydrogen. About 200 thermochemical cycles have been identified but the copper–chlorine thermochemical cycle is regarded as one of the most promising approaches. The Cu-Cl thermochemical cycles are of different types. The numbers of steps (3-steps, 4-steps or 5-steps) have influences on the scale-up challenges and overall cycle efficiencies. A 3-step, 4-step and 5-step cycles of Cu-Cl system for thermochemical water decomposition cycles have been studied extensively [7-9] but there is need for comparative studies that can aid in decision making on the most effective and efficient step cycle especially with respect to the second law efficiency of the process [5].

This study presents the simulation, analysis and parametric studies of the 3 step, 4-setp and 5-step Cu-Cl cycles. It is expected that the study will assist ongoing efforts in seeing to the eventual commercialisation of the process.

The study is presented as follows. In section 2, the detail description of each of the step and the analysis procedures is presented. Section 3 gives the results and discussion and section 4 concludes the study.

## 2 System description, simulation and analysis

### 2.1 Description of the 3-step Cycle

There are two concepts that can be adapted here. Lewis et al. 2009 [10] has three reactions in sequence- Electrolysis, hydrolysis and decomposition. However, the concept of Wang et al. 2009 [11] is adapted for this study and presented in Table 1

**Table 1.** Reactions in the 3-step Cu-Cl cycle

| Step | Reaction | Temperature($^oC$) |
|---|---|---|
| 1 | $2CuCl_2(aq)+H2O(g)\rightarrow 2CuCl(l)+2HCl(g)+1/2O_2(g)+H_2O(g)$ | 400-600 |
| 2 | $4CuCl(s)\rightarrow 2Cu(s)+2CuCl_2(aq)$ | 20-80 |
| 3 | $2Cu(s)+2HCl(g)\rightarrow 2CuCl(l)+H_2(g)$ | 430-475 |

### 2.2 Description of the 4-step cycle

The 4-step cycle combines the processes of hydrolysis of copper and drying step in the 5-step cycle, to produce hydrogen gas and molten CuCl. The 4-step cycle combines these processes together to eliminate the intermediate production and handling of copper solids, through a CuCl/HCl electrolyzer that produces hydrogen electrolytically and aqueous Cu(II) chloride and it has an advantage of requiring less equipment.

The chemical reactions involved in the 4-step Cu-Cl water decomposition as adapted from Wang et al. 2009 [11] are described in the Table 2.

**Table 2.** Reactions in the 4-step Cu-Cl Cycle

| Step | Name | Reaction | Conditions |
|---|---|---|---|
| 1 | Hydrolysis | $2CuCl_{2(aq)}+2H_2O_{(g)}\rightarrow Cu_2OCl_{2(s)}+2HCl_{(g)}+H_2O_{(g)}$ | 375 - 400$^0$C |
| 2 | $O_2$ Production | $Cu_2OCl_{2(s)}\rightarrow 1/2O_{2(g)}+2CuCl_{(l)}$ | 500 - 530$^0$C |
| 3 | Electrolyzer | $4CuCl_{(s)}\rightarrow 2Cu_{(s)}+2CuCl_{2(aq)}$ | 30 -80$^0$C |
| 4 | $H_2$ Production | $2Cu_{(s)}+2HCl_{(g)}\rightarrow 2CuCl_{(l)}+H_{2(g)}$ | 430 -475$^0$C |

### 2.3 Description of the 5-step cycle

There are 5 steps in this cycle including three thermochemical reactions and one electrochemical reaction. The steps, their reactions and temperature ranges are presented in Table 3.

**Table 3.** The five steps in the Cu-Cl cycle with their corresponding reactions

| Step | Reaction | Temperature Range |
|---|---|---|
| Hydrolysis(S1) | $2CuCl_{2(s)}+H_2O_{(g)}\rightarrow CuO{*}CuCl_{2(s)}+2HCl_{(g)}$ | 375-400°C |
| Decomposition (S2) | $CuO{*}CuCl_{2(s)}\rightarrow 2CuCl_{(l)}+1/2O_{2(g)}$ | 500°C |
| Electrolysis (S3) | $4CuCl_{(s)}+H2O\rightarrow 2CuCl_{2(aq)}+2Cu_{(s)}$ | 25-80°C |
| Drying (S4) | $CuCl_{2(aq)}\rightarrow CuCl_{2(s)}$ | >100°C |
| Hydrogen production (S5) | $2Cu_{(s)}+2HCl_{(g)}\rightarrow 2CuCl_{(l)}+H_{2(g)}$ | 430-475°C |

### 2.3 ASPEN-HYSYS Simulation

HYSYS is a process modeling package typically used for steady state simulation, performance monitoring, design, control and optimization of various chemical processes. It predicts the behaviour of chemical reactions using relationships such as mass and energy balances, equilibrium relationships and rate correlations (i.e. reaction and mass/heat transfer). By choosing the appropriate unit operations and thermodynamic models, reliable thermodynamic data and realistic operating conditions, Aspen Hysys uses mathematical models to predict the performance of the cycle and actual plant behavior. The cycles were simulated in HYSYS. The simulation steps are adding appropriate component list, fluid package selection and defining reactions in the basis environment while the streams and unit operations were specified and defined in the simulation environment. The simulation of the steps were carried out with experimental data from previous researches [10-12] and thermodynamic data from literatures.

### 2.3 Thermodynamic analysis

The physical Exergy of the stream was calculated using:

$$E_{xphy} = \Delta H\text{-}T_o\Delta S \tag{1}$$

Where, $\Delta H$ is the change in enthalpy, $\Delta S$ is the change in entropy, $T_o$ is the reference temperature.

$$\Delta H = A\times T + \frac{T^2}{2} + C\times\frac{T^3}{3} + D\times\frac{T^4}{4}\text{-}E\times\frac{1}{T} + F\text{-}H \tag{2}$$

$$S = A\times lnT + B\times T + C\times\frac{T^2}{2} + D\times\frac{T^3}{3}\text{-}E\times\frac{1}{2T} + G \tag{3}$$

$$\Delta S = S\text{-}S_0 \tag{4}$$

Where T is 1/1000 of the specified temperature (in K) of the component and A, B, C, D, E, F, G and H are constants.
The chemical exergy is calculated as

$$E_{ch} = (\mu^0\text{-}\mu_0^0) + RT_0 \ln\left(\frac{c}{c_0}\right) \tag{5}$$

$$E_{heat} = \left(1 - \frac{T_0}{T}\right)Q \tag{6}$$

The total exergy of a stream: The total exergy into and out of a unit is calculated using the following equation;

$$E_X = E_{phy} + E_{ch} + E_{heat} \tag{7}$$

The exergy efficiency of each process step is calculated as

$$n = \frac{Total\ output\ exergy}{Total\ input\ exergy} \times 100 \ldots\ldots\ldots\ldots\ldots\ldots\ldots\ldots \tag{8}$$

## 3 Results and discussion

The energy efficiency, exergy efficiency and irreversibilities of all steps in each cycle and the overall cycle was conducted. Parametric studies were also conducted for each cycle and the corresponding efficiencies were evaluated. Fig.1 gives the HYSYS simulation of the 3-step cycle.

Energy and exergy efficiency of the five-step Cu-Cl cycle is calculated to be 35.89% and 59.64%, respectively. Exergy efficiency increases with the increase in the reaction temperature of both the hydrogen production and drying steps reactor but decreases in the hydrolysis step as seen in fig. 2. Also optimum temperatures for the exergy efficiency for each step in the cycle was found to be 300$^0$C, 100$^0$C, 20$^0$C 100$^0$C and 350$^0$C for the hydrolysis step , decomposition step, electrolysis step , drying step and hydrogen production step respectively. The exergy efficiency of the cycle is seen to increase with increasing reference temperature too.

The overall energy and exergy efficiencies of the four-step CuCl cycle are obtained as 49.39% and 44.78%, respectively. In a related work, the exergy efficiency was found to be 42% [13] and the energy efficiency was 21.6% [14]. The overall efficiency decreases with increasing reference environment temperature from -5$^0$C to 30$^0$C.
The overall efficiency increases with increased temperature of the hydrolysis reaction step. The exergy efficiency of the hydrolysis and the hydrogen production steps of the cycle increases with increase in temperature as seen in fig. 3.

The overall exergy efficiency of the three-step Cu-Cl cycle is 78.5%. There is 0.03% decrease in the overall exergy efficiency of electrolysis reaction step as temperature is increased. Variation of feed rate of $CuCl_2$ and $H_2O$ at 2:1 results in exergy efficiency that ranges from 73.9% to 86.2% while ratio 1:1 ranges from 41.96% to 26.66% as seen in fig.4.

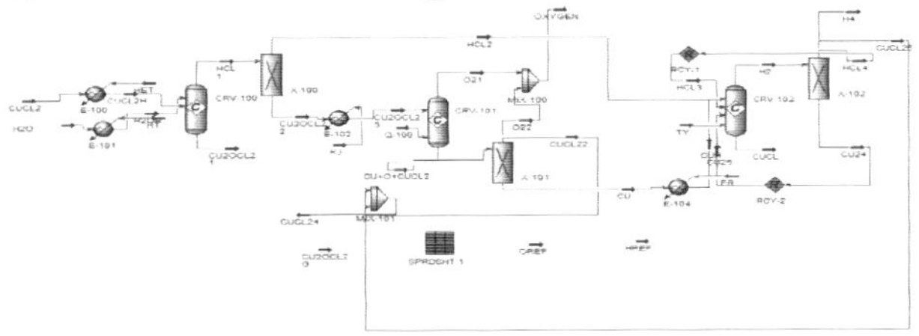

**Fig. 1.** HYSYS Simulation of the 3-step cycle

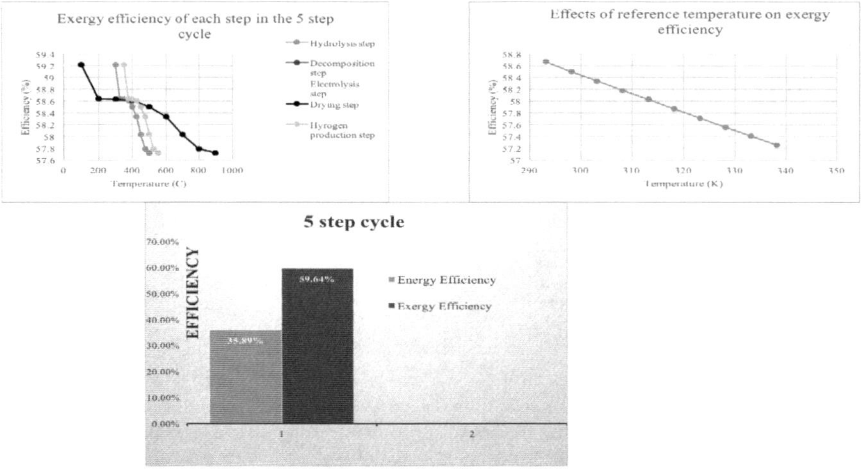


**Fig. 2**. Parametric studies of the 5-step cycle

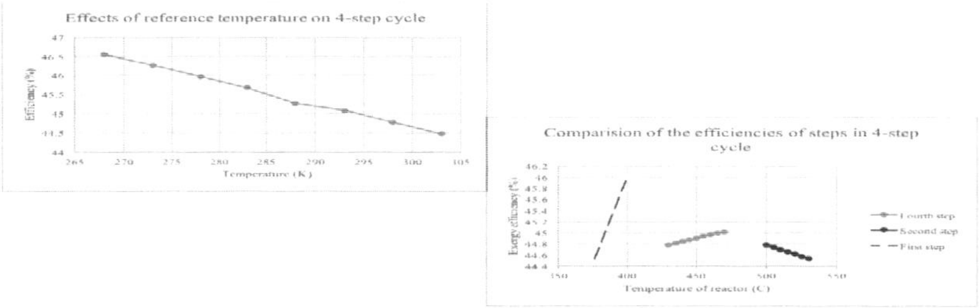


**Fig. 3.** Parameric studies of the 4-step cycle

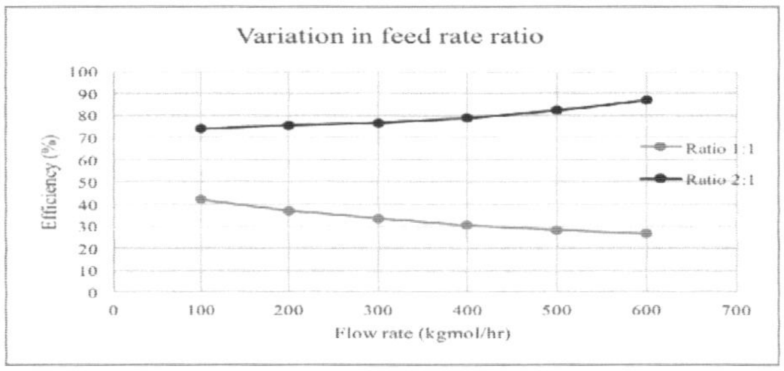


**Fig. 4**. Parametric study of the 3-step cycle

## 4 Conclusion

- The results of the exergy analysis gave 59.64%, 44.74% and 78.21% for the 5-step cycle, 4-step cycle and 3-step cycle respectively while that of the energy analysis is 50%, 49% and 35% for the 5-step cycle, 4-step cycle and 3-step cycle respectively.

- Parametric studies were conducted and possible exergy efficiency improvement of the cycles were found to be between 59.57-59.67%, 44.32-45.67% and 23.50-86.2% for the 5-step, 4-step and 3-step respectively.

- The 3-step cycle seems to be the most energy efficient in terms of second law analysis when compared to other steps cycles. It also seems to be the most receptive to parametric variations.

- The most challenging step in the cycle is the electrolyzer design. Simulation of this step will require themodynamic models that will ably depict the actual electrolysis process.

## References

1. M.F. Orhan, I. Dincer, M.A. Rosen, Int. J. Energy Res, **37**, (2013)

2. M.A. Lewis, M.S. Ferrandon, D.F. Tatterson, P. Mathias, Int. J. Hydrogen Energy, **34**, (2009)
3. A. Hemschemeier, A. Melis, T. Happe,(2009), Photosynthesis Res., **102**, 2-3, (2009)
4. M. A. Laguna-Bercero, Int Journal of Hydrogen, **11**, (2011)
5. G.F. Naterer, S. Suppiah, L. Stolberg, M. Lewis, Z. Wang, M.A. Rosen, I. Dincer, K. Gabriel, A. Odukoya, E. Secnik, E.B. Easton, V. Papangelakis, Int. J. Hydrogen Energy, **40**, (2015)
6. M.T. Balta. I. Dincer, A. Hepbasli, Int. J. Hydrogen Energy, **41,** (2016)
7. D. Yadaf, R. Banerjee, Renew. Sust. Ener. Rev. **54**, (2016)
8. W. Wu, H.Y. Chen, J. Hwang, Int. J. Hydrogen Energy, **42** (2017)
9. M.L. Villeta, M. Costa, N. Massarotti, Renew. Sust. Ener. Rev. **74** (2017) 71-88.
10. M. A. Lewis, M. S. Ferrandon, D. F. Tatterson, P. Mathias, Int. J. Hydrogen Energy, **34**, (2009)
11. Z. L. Wang, G. F. Naterer, K. S. Gabriel, R. Gravelsins, V. N. Daggupati, *Int. J. of* Hydrogen Energy, **34**, (2009)
12. M.A. Rosen, G.F. Naterer, C.C. Chukwu, R. Sadhankar, S. Suppiah, Int. J. Energy. Res., **36**, (2012)
13. M. F. Orhan, D. Ibrahim, A. R. Marc Int. J. Hydrogen Energy, **36**, (2011)
14. M. B. Tolga, D. Ibrahim, H. Arif Energy , **35**, (2010)

# Nowcasting the Output Power of PV Systems

*Marius* Paulescu[1,*], *Oana* Mares[2], *Ciprian* Dughir[2], and *Eugenia* Paulescu[1]

[1]Faculty of Physics, West University of Timisoara, V Parvan 4, 300223 Timisoara, Romania
[2]Department of Physical Foundations of Engineering, Politehnica University Timisoara, V Parvan 2, 300223 Timisoara, Romania
[3]3Electronics and Telecommunications Faculty, Politehnica University Timisoara, V Parvan 2, 300223 Timisoara, Romania

**Abstract.** This paper presents an innovative procedure for nowcasting the energy production of PV systems. The procedure is relayed on a new version of two-state model for forecasting solar irradiance at ground level and a simplified description of the PV system. The results of testing the proposed procedure against on field measured data are discussed. Generally, the proposed procedure demonstrates a better performance than the main competitor based on ARIMA forecasting of the clearness index.

## 1 Introduction

Unlike the power generated by the traditional power plants based on fossil or nuclear fuels, the output power of the photovoltaic (PV) plants is highly variable to erratic. This is due to passing clouds which may cause a fast variation of the solar irradiance inducing further a massive shift in the output power of a PV plant [1]. In this context, forecasting the power generated by PV plants becomes a critical task for smart management of the grid. Accurate forecasts will enable the computers to take control actions aiming to balance the power grid in real-time. Since there are circumstances when the fluctuations of the solar irradiance are on a time scale of minutes or less, nowcasting the output power of PV plants becomes a timely research area [1-3].

This paper presents a new procedure for nowcasting the output power of a PV plant. The procedure involves two interrelated issues: (1) forecasting the solar resource availability and (2) realistic modelling of the PV converter response. The basic concept of the proposed procedure is based on several innovative results obtained by our team: (1) a new procedure of solving the five-parameter model of a solar cell [4], (2) accurate modelling of the PV module operation in real environmental conditions [5] and (3) the innovative two-state model for short-term forecasting solar irradiance [6] based on sunshine number (*SSN*), a binary quantity stating whether the Sun is shining or not.

The paper is organized as follows. Section 2 is devoted to description of the relevant data. The proposed procedure is introduced in Sec. 3. Since the accuracy of nowcasting the output power of a PV system is conditioned by the accuracy of nowcasting the solar irradiance, Sec. 3. is mainly focused on this issue. The results of testing the proposed procedure against measured data are discussed in Sec.4.

* Corresponding author: marius.paulescu@e-uvt.ro

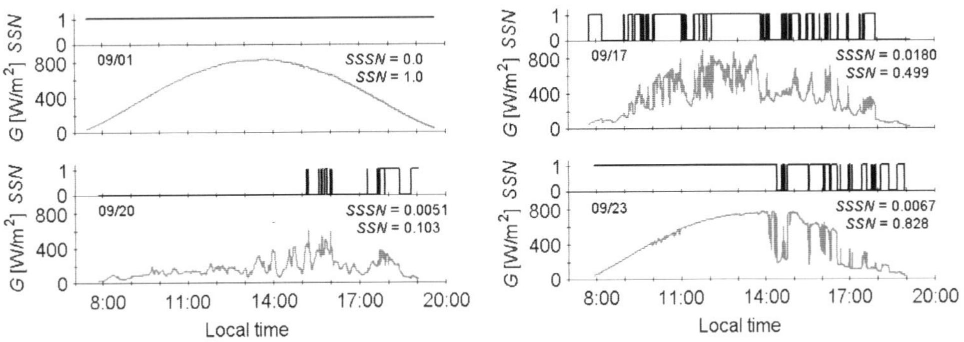


**Fig. 1.** Variation in time of the global solar irradiance (*G*) and sunshine number (*SSN*) recorded in four different days of 2016. The average values of *SSN* and *SSSN* are displayed on each graph.

## 2 Data

The study was conducted with data recorded on the Solar Platform of the West University of Timisoara, Romania (http://solar.physics.uvt.ro/srms). The town of Timisoara (45°46′N; 21°25′E and 85 m asl) has a warm temperate climate, typical for the Pannonia Basin (Köppen climate classification *Cfb*). The built database contains radiometric, meteorological and electrical data collected from an experimental PV setup. All the sensors are integrated into an acquisition data system based on National Instruments PXI Platform and explored simultaneously at high temporal resolution (15 seconds). The global and diffuse solar irradiance are measured using DeltaOHM LP PYRA 02 first class pyranometers which fully comply with ISO 9060 standard and meet the requirements defined by the World Meteorological Organization (WMO).

### 2.1 Post-processed data

Mathematically, the sunshine number is defined as a time dependent random binary variable, as follows [7]:

$$SSN_t = \begin{cases} 0 & \text{if the sun is covered by clouds at time } t \\ 1 & \text{otherwise} \end{cases} \tag{1}$$

Series of *SSN* values have been derived using the WMO sunshine criterion [8]: the Sun is shining at time *t* if the direct solar irradiance exceeds 120 W/m$^2$. Therefore:

$$SSN_t = \begin{cases} 1 & \text{IF } \left(G_t - G_{d,t}\right)/\sin h_t > 120 \text{ W/m}^2 \\ 0 & \text{otherwise} \end{cases} \tag{2}$$

where $G_t$ and $G_{d,t}$ denote the measured global and diffuse solar irradiance at the time *t* and $h_t$ is the Sun elevation angle. On basis of *SSN* a straightforward quantifier for the variability of the solar radiative regime is defined in Ref. [9]. This is the sunshine stability number:

$$SSSN_t \equiv \begin{cases} 1 \text{ if } SSN_t > SSN_{t-1} \\ 0 \text{ otherwise} \end{cases} \tag{3}$$

The average value of *SSSN* during a time interval $\Delta t$, $\overline{SSSN}$, ranges between 0 and 1/2, measuring the frequency of *SSN* the changes during $\Delta t$. Daily values of $\overline{SSSN}$ have been calculated in this study. Figure 1 illustrates the time series of $G$ and $SSN$ in four days of 2016. In every day, more than 2500 records are used to build the graph.

Another post-processed quantity is clearness index [10]:

$$kt_t = G_t / G_{e,t} \tag{4}$$

where $G_t$ and $G_{e,t}$ are the global solar irradiance measured at ground level and its counterpart estimated at the top of the atmosphere. Clearness index isolates the stochastic components in solar irradiance time series. The traditional approach in nowcasting solar irradiance is based on the statistical modeling of the clearness index time series.

## 3 Model

The proposed procedure for nowcasting the output power of a PV system runs two algorithms: (1) nowcasting the solar irradiance and (2) modelling the PV system. Both algorithms are briefly described next.

### 3.1 The 2-state model v.2

In this study an upgraded version of the two-state model for nowcasting solar irradiance [6] is tested. The new version of the model connects an empirical model for estimating the clear sky solar irradiance with a statistical model for forecasting $SSN$:

$$\hat{G}_t = \begin{cases} c_{cs} \cdot G_{0,t} & \text{if} \quad S\hat{S}N_t = 1 \\ \overline{\tau}_c \cdot G_{0,t} & \text{if} \quad S\hat{S}N_t = 0 \end{cases} \tag{5}$$

where $\hat{G}_t$ is the forecasted value of the solar irradiance at time $t$. $G_{0,t}$ is the estimated solar irradiance under clear sky at time $t$ $c_{cs}$ is a dynamic correction applied to the mean atmospheric transmittance encapsulated into the $G_{0,t}$ equation. $\overline{\tau}_c$ is the attenuation applied to $G_{0,t}$ according to the cloud transmittance. Note that the sole parameter that needs to be forecasted in Eq. (5) is $SSN$. The model operates as follows ( $S\hat{S}N_t$ denotes the forecasted value of $SSN$ at time $t$). On the state $S\hat{S}N_t = 1$, the global solar irradiance is being estimated using a clear-sky model. In the first version of the two-state model $c_{cs}$ was assumed equal to one. On the state $S\hat{S}N_t = 0$, the clear-sky estimate is adjusted with the cloud transmittance $\overline{\tau}_c$. Both parameters $c_{cs}$ and $\overline{\tau}_c$ are estimated simultaneously by a linear regression applied to data measured in a period $\Delta t$ prior to the forecasting moment. Regarding $\Delta t$, an important finding was reported in [11]: the minimum error is reached if $\Delta t$ equals the forecasting lead time.

### 3.2 PV convertor

The algorithm for modelling the PV converter combines two results obtained by our team. Firstly, the PV module current-voltage characteristics at STC is evaluated in terms of one-diode model. The model parameters are extracted using the simple procedure from Ref. [4]. Secondly, the procedure from [5] is applied for translating the I-V curve from STC to an arbitrary operating point defined by the in-plane solar irradiance and the solar cell temperature.

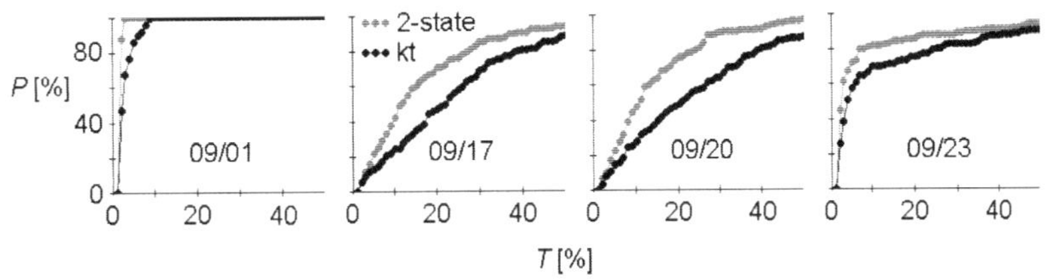


**Fig. 2.** Percentage of forecasts *P* accurate to within a given tolerance interval *T* centered on the measurements. Lead time = 5 minutes.

**Table 1** Models performance in nowcasting solar irradiance. *P* (*T* = 5%) is the percentage of forecasts accurate to within *T* = 5%. *N* represents the total number of measurements. Lead time = 5 min.

| **Date** | **09//01** | | **09/17** | | **09/20** | | **09/23** | |
|---|---|---|---|---|---|---|---|---|
| *N* | 2932 | | 2736 | | 2697 | | 2658 | |
| Model | 2-state | kt | 2-state | kt | 2-state | kt | 2-state | kt |
| *nMBE* | 0.000 | -0.006 | 0.001 | 0.005 | -0.001 | -0.016 | 0.001 | 0.005 |
| *nRMSE* | 0.003 | 0.012 | 0.258 | 0.328 | 0.227 | 0.367 | 0.201 | 0.222 |
| *P* (*T* = 5%) | 1 | 0.895 | 0.251 | 0.142 | 0.278 | 0.160 | 0.732 | 0.604 |

# 4 Results and discussions

The forecasting model was tested on the Solar Platform against data measured on a fully monitored PV module directly connected to a pure resistive load. The model performance is illustrated on basis of data recorded in four days from 2016. As Fig. 1 indicates, these days were characterized by different solar radiative regimes, ranging from a stable one (day 09/21) to a high unstable one (day 09/17).

## 4.1 Nowcasting solar irradiance

Based on the results from [6], the ARIMA(0,1,0) model was used for nowcasting *SSN*. The clear-sky solar irradiance ($G_{0,t}$ in Eq. 5) was estimated with the Biga and Rosa's [12] empirical model. The cloud transmittance $\overline{\tau}_c$ was estimated as follow. As stated in Sec. 3, on the Solar Platform data are recorded 15 s time interval. In every day, at the moment when the sun elevation angle reaches 5°, the cloud transmittance is being initialized to a mean value $\overline{\tau}_c = 0.29$. Further during the day, $\overline{\tau}_c$ is periodically updated. At each forecasting moment, the previous values of *SSN* are tested to find if there were a moment with *SSN* = 0. If yes, $\overline{\tau}_c$ is updated. If not, the last value $\overline{\tau}_c$ is kept. In this way, $\overline{\tau}_c$ is always estimated over the most recent measurements and the most suitable value is provided in Eq. (5). A similar procedure was applied to estimate $c_{cs}$, but using $c_{cs} = 1$ as seed in the early morning.

In a second stage of the test, the performance of the 2-state model v.2 was compared to the one of a standard solar irradiance forecasting model. The ARIMA model applied to the clearness index series is considered as one of the main competitors. The statistical software Statgraphics was employed for fitting the ARIMA models on *kt* data series According to AIC criterion, the model selected for forecasting *kt* was ARIMA(2,1,2). The performances of the two-state v.2 model and *kt*-based model are compared in Table 1. Considering any statistical indicator the two-state model performs more accurate than the *kt*-based model.

 https://doi.org/10.1051/e3sconf/20186100010

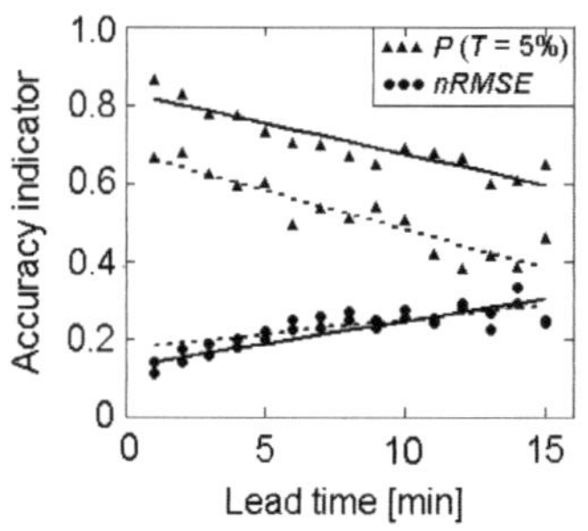


**Fig. 3.** Accuracy of nowcasting solar irradiance by the two-state model (line) and *kt*-based model (dashed) in 09/23 with respect to the lead time.

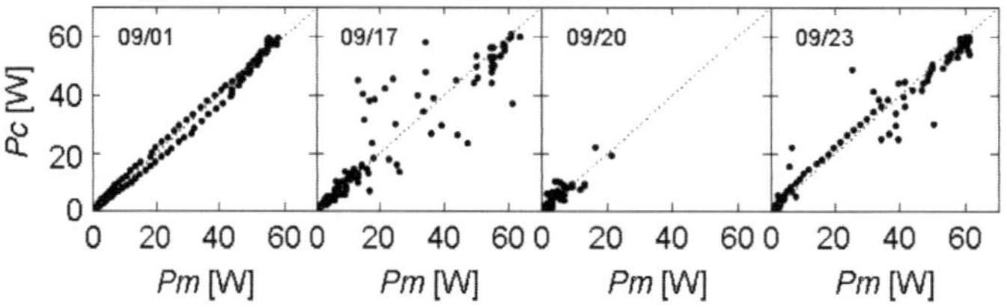


**Fig. 4.** Accuracy of nowcasting the output power of the monitored PV system. Lead time = 5 minutes.

If the percentage of forecasts accurate to within a given interval is assessed, a remarkable advantage of the two-state model is noticed. This is more visible in Fig. 2, where the percentage $P$ of forecasts accurate to a given interval tolerance interval $T$ is plotted against $T$, expressed in percent. In general, for T < 20%, the percentage of forecasts accurate to within $T$ is far greater in the case of the two-state model. At $T$ = 40% the success rates of the two models become comparable, but the two-state model still preserves an advance. It is worth to note that the two-state model performs accurate even in a very unstable solar radiative regime (Fig. 2, day 09/17).

Figure 3 shows the models performance in respect to the forecasting lead time in 09/23, a day with a stable solar radiative regime in the morning and high variable one afternoon. Visual inspection shows that while the two models exhibit the same accuracy (measured by *nRMSE*), a remarkably higher precision (measured by $P$) of the two-state model is noted. For the entire range of lead time considered in the illustration from Fig. 3, the two-state model precision $P(T = 5\%)$ exceeds with 20% the precision of the *kt*-based model.

### 4.2 Nowcasting the PV output power

Figure 4 assesses graphically the accuracy of the proposed procedure in nowcasting the output power of the monitored PV system for a lead time of 5 minutes. The performance is high in 09/01, a clear-sky day characterized by a stable solar radiative regime $\left(\overline{SSSN} = 0\right)$. It decreases with the increasing of cloudiness and the instability of the solar radiative regime. The lowest performance is noted in 09/17, a day characterized by a high unstable solar radiative regime $\left(\overline{SSSN} = 0.018\right)$. This behavior is induced by the decreasing of the ARIMA model accuracy in nowcasting *SSN* with the increasing of the instability of the solar radiative regime. The cause is the model inertia, tending to preserve the current state of *SSN* [13]. Although many efforts were devoted for improving the accuracy of nowcasting sunshine number [3], further research for removing the persistence are required.

## 5 Conclusions

In this study an innovative procedure for nowcasting the output power of a PV system was reported. This is based on an upgraded version of the two-state model for nowcasting solar irradiance. Overall results show that the two-state model performs better in nowcasting solar irradiance than the main competitor, ARIMA forecasting of the clearness index. Always when sunshine number is correctly predicted, the forecasts trace with accuracy the measurements. This give a significative advantage to the two-state model when the percentage of the forecasts accurate to within a given tolerance interval is evaluated. A notable advantage of the two-state model is its modularity, which allows us to increase the forecasting accuracy by fine tuning its components: the clear-sky model, the dynamic adjustment of the atmospheric transmittance and the model for nowcasting sunshine number. The proposed procedure provides high-quality forecasts of the PV output power when high and moderate values of solar irradiance are recorded. Further research for improving the forecasting accuracy in low irradiance conditions and/or high unstable solar radiative regime are required.

## Acknowledgements

This work was supported by a grant of the Romanian National Authority for Scientific Research and Innovation, project number PN-III-P2-2.1-PED-2016-0592.

## References

1. A. Mills, M. Ahlstrom, M. Brower, A. Ellis, R. George, T. Hoff, B. Kroposki, C. Lenox, N. Miller, M. Milligan, J. Stein, Y. Wan. IEEE Power Energy M. **9**, 33 (2011).
2. M. Brabec, M. Paulescu, V. Badescu. Sol Energy **101**, 272 (2014).
3. Y. Chu, H.T.C.Pedro, M. Li, C.F.M.Coimbra. Sol Energy **114**, 91 (2015).
4. O. Mares, M. Paulescu, V. Badescu V. Energ Convers Manage **105**, 139 (2015).
5. M. Paulescu, V. Badescu, C. Dughir. Energy **70**, 49 (2014).
6. M. Paulescu M, O. Mares, E. Paulescu, N. Stefu, A. Pacurar, D. Calinoiu, P. Gravila, N. Pop, R. Boata. Energ Convers Manage **79**, 690 (2014).
7. V. Badescu, M. Paulescu. Atmos Res **101**, 194 (2011).
8. World Meteorological Organization. *Guide to Meteorological Instruments and Methods of Observation*. WMO-No.8/2008.
9. M. Paulescu, V. Badescu. Theor Appl Climatol **103**, 459 (2011).
10. B.Y. Liu, R.C. Jordan. Sol Energy **4**, 1 (1960).
11. D. Costa, O. Mares. AIP Conference Proceedings **1634**, 151 (2014).
12. A.J. Biga, R. Rosa. Sol. Energy **23**, 61 (1979).
13. M. Paulescu, V. Badescu, M. Brabec. Energy **54**, 104 (2013).

# Hybrid Renewable Power Systems for Generation of Own Power by Small and Medium-Scale Enterprises

*Archishman* Bose[1,*], *Kiran* Raj[1] , *Denitsa* Kuzeva[1], *Tommaso* Mura[1], *Jialei* Xin[1], Jeevan Jayasuriya[1,2]

[1]InnoEnergy Master School, MSc. SELECT, InnoEnergy Scandinavia, 114 28, Stockholm, Sweden
[2]KTH Royal Institute of Technology, Department of Energy Technology, 100 44, Stockholm, Sweden

**Abstract.** Decentralized power generation, from renewables, is an attractive option for the future energy transition. Through a case study, the techno-economic feasibility to produce own power from distributed renewable to de-carbonize the operations of the Small and Medium Scale Enterprises (SMEs) was critically analysed. The case study was performed on one of the leading printing outfits of Sri Lanka. Solar photovoltaic (PV) and biomass gasification systems are the most cost-efficient and easy to operate technologies for grid-connected, small-scale power generation, at present, for the context. Grid integration has been found as a major challenge, in both technical and economic parameters of the project. The low capacity factor of solar PV and complexity of the supply chain for biomass power systems are critical to the respective technologies. A hybrid Solar PV-Biomass gasification power plant would have superior techno-economic performances with lower environmental impact than stand-alone systems. An equal share of the net power capacity between the technologies was obtained as the most suitable combination for the proposed hybrid power plant. A net carbon dioxide reduction of more than eighty percent of the operations of the SMEs is feasible. Socio-political factors also have a high impact on overall viability of such small-scale systems.

## 1 Introduction

Technological progress, along with human development is leading to an increasing consumption and demand of energy [1]. Besides, unplanned growth, together with the unregulated use of fossil fuels has resulted in significant acceleration of climate and environmental degradations observed and described by the scientific community [2]. From the consumption point of view, industries accounted for 42% of the world final energy consumption in 2015 [4]. Small and Medium scale enterprises (SMEs), comprising most of the industrial sector [5], especially in the developing countries, accounts for a high share of the total industrial energy consumption. Apart from energy efficiency measures, own

* Corresponding author: arcbos@kth.se

energy generation from renewables forms an innovative strategy to decrease the carbon footprint of operations of such industries.

Power generation from renewables is a major instrument towards promoting the production of clean energy and tackling climate change. Distributed power generation, due to multiple advantages, will play a crucial role in this transformation, besides centralized power systems, [7]. However, technological challenges and cost-competitiveness of renewable energy-based power production, both with respect to grid integrated and off-grid systems, are major challenges, especially in developing countries. Lack of adequate technical knowledge, especially for the development of small-scale power generation systems often lead to poor system design resulting in economic failure. A strong technological framework and support, together with national and international financial strategies like subsidies, public funds and innovative policies are crucial as well for the development of distributed renewable power generation units.

However, generation of own green power has its own drawbacks. Often the SMEs would lack adequate technical knowledge for selection and design of such systems. Over-simplified evaluations based on economic assessments, though often attractive, include inappropriate assumptions, resulting in poor system design and economic failure of such systems. Besides SMEs are usually located either within urban areas or have inadequate space within their premises, presenting considerable barriers to set up such power generation units on site.

Thus, in the present case study, the techno-economic feasibility of an offsite, small-scale grid integrated renewable power generation system for own energy needs by SMEs was studied considering multiple perspectives technology availability, economic feasibility, as well as social, environmental and policy implications.

## 2 The Case Study

The necessary evaluations were performed through a case study. Accordingly, a leading newspaper company in Sri Lanka, representing a typical SME, was selected to evaluate the feasibility of a small-scale grid-connected renewable power plant to decarbonize its operations. The use of electricity for printing and the presence of fossil-derived power plants in the electricity mix of Sri Lanka [6] results in inherent carbon emissions from printing operations. The annual electricity consumption was estimated as 5.2 GWh.

The newspaper printing facility is in a largely urbanized location near Colombo, Sri Lanka. An off-site grid integrated power generation system was therefore considered to be analysed. Due to high land prices and local land acquiring policies, around 40.5 hectares (ha) of land in Bingiriya, in the North-Western Province of Sri Lanka, 120 km distance from the printing facility was identified as suitable. The rural area is in the coconut belt, within the dry zone of Sri Lanka.

The available land comprises a tank, 12.2 ha of paddy fields and around 28.3 ha of coconut plantations. The river Deduru flows within 1000 metres from the land. However, an upstream dam results in the river to run dry most of the year, raising water constraint issues for the project. Other analyses of roads, infrastructure, site vegetation, soil condition, surrounding locality including cultural and environmental aspects were conducted for the overall feasibility of a renewable energy project development.

### 2.1 Assumptions

Several rational assumptions were made including a conservative evaluation approach. Especially, for a biomass power generation unit, a well-designed and maintained biomass-based power generation unit can achieve around and above 70% capacity factor for

yearlong operations [7]. However, a capacity factor of 60% for the present evaluations was considered along with inferior overall system efficiency of 20%.
Besides technological feasibility assessment, capital expenditures (CAPEX), operation expenditures (OPEX), project internal rate of return (IRR), payback period and net present value (NPV) were evaluated. Project lifetime of 20 years was assumed. A debt-equity ratio of 70:30 was selected for financial evaluations. Grid balancing fees and carbon credits were not considered based on local laws. Only locally available and technologically matured solutions were considered for system design due to ease of project implementation and operation. Detailed economic parameters were assumed based on the local Sri Lankan policies as follows:

Corporate tax: 10%; Capital insurance rate: 0.50%; Decommissioning factor: 5%; Inflation rate in Sri Lanka: 9.72%; Cost of Equity 16%; Tax holiday period: 6 years [8]. Cost of Biomass Feedstock: 40 $/tonne

Considering limitations of SMEs, capital expenditure (CAPEX) was limited to 3 million $ considering a small-scale project. A payback period of around five years with a project internal rate of return (IRR) of over 10% was considered for the economic feasibility of such small-scale projects.

## 3 The Case Study

### 3.1 Grid Connection

Grid connection, especially for fluctuating renewables presents technical challenges [9]. The nearest 132/33 kV Grid Substation, being 30 km distance away, a severe grid integration challenge was faced. The necessary infrastructure of a dedicated medium voltage line would cost an estimated 0.6 million $ severely affecting the economic viability of the small power generation system.

A 33KV Medium Voltage feeder was assessed to be feasible for grid connection based on the carrying capacity of the feeder. A 33KV feeder, connecting two substations, around one kilometre from the selected land was identified to a have a feed-in capacity of 3MW. The absence of power generation units connected to the feeder would allow the grid integration of the proposed unit at a cost of nearly fifty times less, at around 12000 $. However, with an increase in distributed power generation units, grid integration will become a major techno-economic challenge for a small-scale grid-connected power generation unit.

### 3.2 Resource Feasibility and System Design

Solar, wind and biomass-based power generation are the most mature forms of renewable power generation for small to medium scale applications.

#### *3.2.1 Wind*

On-shore wind turbines are currently the most economical option for harnessing wind energy on a small scale. Even though Sri Lanka has an excellent wind potential, the identified site is in an area with poor or marginal wind potential [10] . Additionally, the selected land is surrounded by coconut plantations. Being over 10m high, this would severely impact the performance of wind turbines. Besides, the area is frequented by migratory birds, presenting significant social and environmental negativities towards developing a wind farm. Unfavourable resource, high land requirement, together with a lower plant factor would limit small-scale wind power units.

### 3.2.2 Solar

Among different technologies to harness solar energy, solar photovoltaics is one of the most promising and fastest growing industries worldwide [11]. The yearly average solar insolation was estimated at around 1600-1700 kWh/m2 with a yearly mean daily average ambient daytime temperature ranging from 24-30 $^{o}$C at the selected location. A solar PV generation system was therefore technologically viable.

A 4MW standalone solar PV power plant with an average annual capacity factor of 17.5% would be necessary to produce 5.2 GWh annual energy. The net land requirement would be around 80 ha.

However, as discussed, the grid connection feasibility is limited to 3MW without major economic constraints for the selected case study. Hence a stand-alone 3MW solar PV power plant would fail to meet the necessary generation requirements of 5.2 GWh per year for the printing operations.

### 3.2.3 Biomass

Even though medium to large scale biomass to power is a matured technology, the technology development issues remain regarding small power plants (around the order of 1 MW) [7].

The primary limitation to the development of biomass power plant is the availability of resources, namely raw biomass and water. Any capacities over 2MW, as studied, would require extensive resource management, or over-dependence on external supplies, often leading to unwanted complicacy of operations of small-scale projects. Gasification-based systems are preferable for small-scale operations. Modular, engine-based gasification units were found to be the most suitable technology for small-scale biomass to electricity applications. Besides requiring less space with lower water consumption, the presence of internal combustion engines results in a more robust system with higher part load efficiencies [12].

Studied feedstocks included energy crops and agricultural wastes. Co-cultivation of dendro (Gliricidia sepium) and coconut was obtained as the optimum solution for establishing a dedicated supply chain for the proposed biomass power plant in Sri Lanka. Along with dendro, an energy crop, coconut wastes, including shells and husks was considered as the primary feedstock. Based on local studies conducted by the authors, a mixture of 39% by weight coconut wastes and 61% by weight of dendro is feasible to be obtained through co-cultivation. A considerable LHV of 16 MJ/kg of the mixture (20% moisture) would result [13]. Availability of raw material from within the material flow of the SMEs was considered as a valuable resource. Waste paper, currently sold for recycling was found as an economically feasible resource from within the printing facility, securing the supply chain further.

Segregated solid fractions of municipal solid wastes (MSW) as an alternative raw material was considered. However, technological and social challenges with regards to collection, sorting and gasification of MSW would render MSW unsuitable as a feed-stock.

A 1MW stand-alone biomass gasification unit, running at 60% capacity factor, would be able to produce the required yearly energy of 5.2 GWh necessary for printing operations.

For such a unit, a feed rate of 24 tonnes/day, would require around 202 ha of dendro-coconut co-cultivated lands. Cost of feedstock would be 40 $ per tonne of mixed dendro-coconut waste feed. 1200 tons of waste paper was identified to be available annually from the newspaper printing facility as waste from printing operations. This would substitute around 20% of the dendro-coconut waste requirement for a 1MW system. The net cost of waste paper, including transportation charges from the printing facility to the biomass power plant, was estimated to be 0.5 Sri Lankan Rupees (LKR)/kg of paper, considerably decreasing the economic burden of the cost of feedstock on the biomass power plant.

An economically optimized strategy to develop dedicated farming and bulk procurement will play a crucial role in the viability and higher capacity operations of the proposed power plant.

#### *3.2.4 Hybrid Power Generation*

Power generation systems from multiple energy sources have significant advantages over those from the single energy source.

Small-scale grid-connected wind energy generation, being obtained both techno-economically challenging, was not considered for a hybrid system design. However, a solar-biomass based hybrid system was found to be technologically feasible in the present case study. Hybridization of the solar photovoltaic plant with the biomass-based power generation unit has been shown to complement each other to produce a stable power overcoming individual drawbacks [14]. Multiple combinations of hybridization based on individual system size were studied, that would be able to supply the required minimum output of 5.2 GWh of annual energy.

Different hybrid systems capacities combining the solar PV and Biomass gasification technologies were considered. Fig. 1. represents the hybridization scheme. The electricity produced by the solar PV power plant is via DC Voltage. Additional inverters are necessary to convert the DC Voltage to AC Voltage for grid integration through a step-up transformer. However, biomass power generation units directly generate AC voltage and hence no additional inverter is required. The two plants would be integrated to the grid by a common AC bus. The net installed capacity of the system would therefore be the sum of the capacities of the individual components of the hybrid system, namely solar PV and Biomass Gasification.

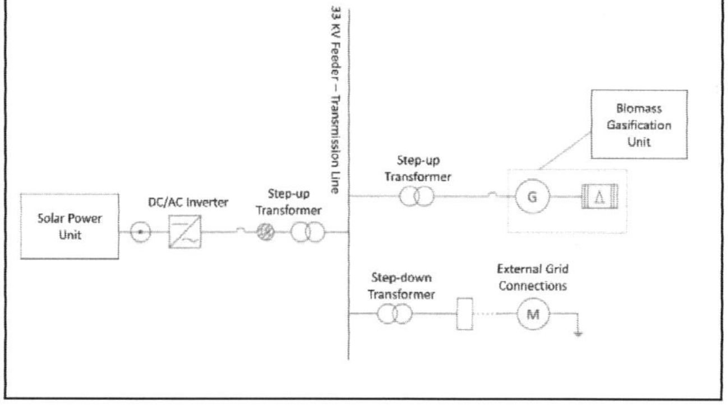


**Fig. 1.** Layout of the 2MW Hybrid System, 1MW each from Solar PV and Biomass Gasification

### 3.3 $CO_2$ Mitigation Potential

Energy generation from onshore wind turbines and solar PV was considered carbon neutral, ignoring lifecycle emissions. Even though biomass to power is considered carbon neutral, net non-negligible emissions result from biomass cultivation, transportation and pre-treatment. Significant emissions with respect to electricity consumption and use of supplementary fossil fuel, occur during construction stage of the project. The net emissions from raw biomass-based power generation unit over a period p (PEp), was calculated based on practical projects [15], according to (1).

$$PE\_p = \{APE\_cul + APE\_pret + APE\_trans\} \times Q\_bio\ ,\_p + ECP\_(J,grid,p) \times EFP\_(J,grid) + ECP\_(J,cap,p) \times EFP\_(J,cap) \quad (1)$$

where, APEcul represents the unit CO2 emissions per tonne of biomass cultivation = 0.0252; APEpret represents the unit project CO2 emissions from fossil fuel and electricity consumption at the various points of the production chain, including pre-treatment facility = 0.0220; APEtrans represents the project specific CO2 emissions from biomass feedstock transportation from different locations = 0.0245; Qbio,p represents the quantity of as-procured biomass (in tonnes); ECPJ,grid,p represents the amount of electricity imported from grid for self-consumption over the project in period p [MWh/p], assumed negligible; EFPJ,grid represents the local grid emission factor = 0.7689 kgCO2/MWh [16]; ECPJ,cap,p represents the amount of off grid electricity consumed by the generator in the project period p [MWh/p], assumed negligible; EFPJ,cap represents the emission factor for captive electricity =0.8 tonneCO2/MWh.

Different feedstocks like waste paper or MSW have significantly different CO2 mitigation potentials, resulting from different system-specific emissions. End of life emissions from waste or read newspapers were calculated based on existing literature [17]. Majority of the emissions results from the release of the energy stored in paper. Hence waste paper to energy was considered as carbon neutral. The net emissions including transport were calculated according to (1) as 0.0245 tonnesCO2/tonne of feedstock.

## 4 Comparative Study

A comparative analysis was developed to assess the performances of the stand-alone and hybrid systems based on the both Solar PV and Biomass Gasification stand-alone power plants and the different combinations of hybrid systems combining the two technologies. Based on assumptions and discussions presented above, five alternative system configurations were considered for the comparative analysis as shown in Table 1. The results are analyzed in the following chapter.

**Table 1** Systems considered for techno-economic comparative evaluations

| | Equivalent System Capacities (MW) | |
|---|---|---|
| **Technology** | **Solar PV** | **Biomass Gasification** |
| System A | 3 | - |
| System B | - | 1 |
| System C | 1 | 1 |
| System D | 0.5 | 1.5 |
| System E | 2 | 0.5 |

## 5 Results

Based on the five systems, a comparative analysis was developed towards obtaining the most suitable solution for a small-scale grid-connected renewable power generation system for production of own green power by SMEs. The results have presented and discussed in the following sections.

### 5.1 Energy Generation Potential

As can be seen from Fig. 2, the net energy generation of the 3 MW Solar PV power plant is 4.6 GWh, which is less than the annual demand of 5.2 GWh by the printing industry, marked by the blue line. Hence a small scale stand-alone solar PV power plant would be unable to meet the demand by itself in the present case study. However, System B, running entirely on biomass generates 5.256 GWh, slightly higher than the energy demand, due to higher plant capacity factor of 60%.

On the other hand, a hybrid system of Solar PV-Biomass Gasification increases the energy generation considerably. It needs to be mentioned here that due to the complicacy of the biomass supply, as discussed in the previous sections, a lower capacity factor of 50% was assumed for the 1.5 MW system. Conversely, the capacity factor of a 500 KW system was assumed to increase to 70% due to lower needs of the feedstock. Hence, the energy generation potential of Systems D and E were evaluated based on the individual biomass power plant capacity factors.

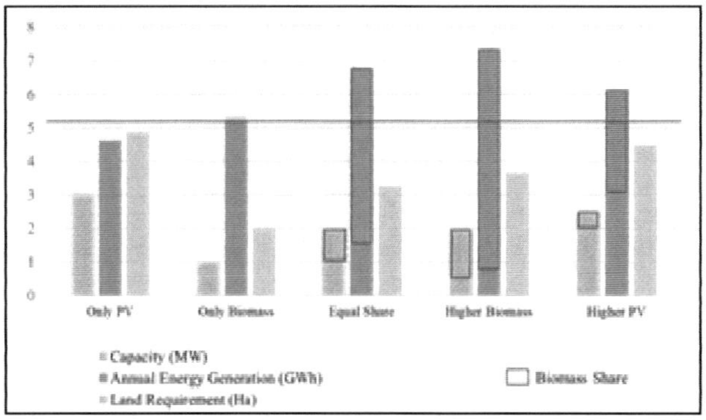


**Fig. 2.** Comparison between System Capacity Net Energy Produced and share of produced energy from biomass of the alternative systems

The capacity factor of the biomass power plant considerably influences the overall energy generation from a solar PV-Biomass Gasification hybrid system. A higher share of solar capacity within the hybrid system considerably decreases the net energy output, as is evident from System E, even though the capacity factor of the 500 kW Biomass power plant was considered higher at 70%.

### 5.2 Economic Comparisons

#### *5.2.1 CAPEX and OPEX*

Table 2 shows the CAPEX and OPEX of the systems considered. The costs of 1MW solar PV and 1MW Biomass Gasification power plants were determined based on quotes from regional manufacturers and local project details. The capital costs of the equipment and infrastructure for the higher and lower capacity systems were assumed based on the six-tenth rule [34]. Due to a rapid drop in prices of PV panels in recent years, the cost of a 3MW solar PV power plant is comparable to a 1MW Biomass Gasification power plant. Also, the upfront costs of civil and additional infrastructure like storage space, cooling towers, etc. are either much lower or non-existent for Solar PV systems than the biomass-based power generation unit. Additionally, biomass gasification power plant, comprising gasifiers, cleaning equipment and internal combustion engines are more complex and costly equipment, increasing the overall CAPEX.

**Table 2** CAPEX and OPEX of the considered systems

| SYSTEM | CAPEX (million $) | OPEX ('000 $/Year) |
|---|---|---|
| System A | 1.72 | 56.01 |
| System B | 1.78 | 304.89 |
| System C | 2.45 | 316.343 |
| System D | 2.88 | 460.06 |
| System E | 2.39 | 193.586 |

Indeed, specific costs of stand-alone systems increase further due to the requirements of dedicated grid connection, electrical systems and administrative infrastructure developments. However, these costs are shared between comprising technologies in a hybrid system, decreasing specific costs of the overall project. The CAPEX of systems D and E hence were determined accordingly.

The system operation costs, including the cost of fuel, is also presented in Table 2. By far, fuel costs increase the overall OPEX of the biomass systems, either stand-alone or hybrid. An increase in capacity of the biomass power plant increases the operating costs of System D. However, the reverse might not always be true. Considering a conservative approach, besides water consumption costs, all other costs were assumed same for a 500KW biomass power unit with regards to that of a 1MW unit.

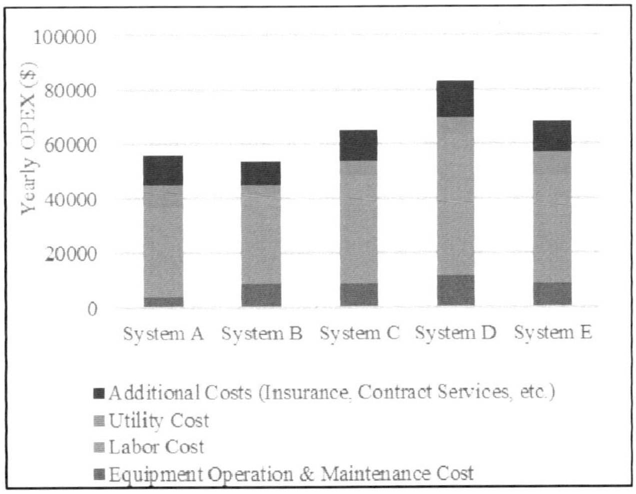


**Fig. 3.** Components of OPEX of the systems studied

Costs of operation and maintenance, that included annual equipment operation and maintenance cost, labour cost, utility costs primarily in terms of water and electricity, and additional costs including insurance, contract services, etc. The labour wages were divided into wages for the plant manager, maintenance personnel and security personnel, based on local salary levels at 12,000 $, 3,200 $ and 2,000 $ per year respectively. The specific costs of operation for stand-alone systems are higher due to requirements of dedicated personnel, besides, plant managers and security. However, for hybrid systems, even though the requirement of dedicated personnel for each system type may increase, the net specific costs for manpower requirements decrease. A breakdown of the OPEX, without the fuel cost of the five different systems considered, is shown in Fig. 3, which could potentially be discussed further as to means of decreasing the OPEX for small-scale power generating units.

### *5.2.2 IRR, NPV, Payback Period and LCOE*

The economic performance of the five systems in terms of Internal Rate of Return (IRR), Net Present Value (NPV), Payback Period (PBP) and Levelized Cost of Electricity (LCOE) was evaluated. The feed-in tariff of biomass was 16.67 c$/kWh, as fixed by government incentives. On the other hand, Solar PV tariff is usually tender based and was assumed as 12 c$/kWh. System C, essentially a 2MW hybrid system of 1MW Solar PV and 1MW Biomass Gasification power generation units are presented in Table 3. Besides an attractive IRR and an acceptable LCOE of 10.4 cents $/KWh, a payback period of 4 years makes the system techno-economically quite attractive.

**Table 3** Economic performance of System C

| Description | Values |
|---|---|
| IRR (%) | 25.4 |
| Net Present Value (million $) | 2.38 |
| Payback Period | 4 |
| LCOE (c$/KWh) | 10.50 |

A comparison of the economic performance of System C with regards to that of the other systems is presented in Fig. 4. As can be clearly seen, the performance of the System B is the poorest. High capital costs, compared with the considerable cost of fuel results in a stand-alone biomass power plant to be less attractive in comparison to a stand-alone solar PV power system, represented by System A. However, all the hybrid systems have superior performance than the stand-alone systems. Even with varying overall system capacities and relative capacities of the component systems, the overall project cash flow of all the three hybrid systems was obtained comparable.

Additionally, the lower feed-in tariff of the solar PV power plant than that of the biomass system results in the IRR of system A to be lowest. This also results in a project payback of 7 years for the solar PV power plant. in comparison to six years for the stand-alone biomass power plant and four to five years of the hybrid systems. Conclusions regarding the economic performance of the hybrid systems are drawn based on Fig. 4. Clearly, the economic performance of System C is superior to Systems D and E. The higher IRR and NPV results in the payback period to be lowest at 4 years of system C. However, the LCOE of the System E is lower due to the relatively higher capacity of the solar PV within the 2.5 MW system.

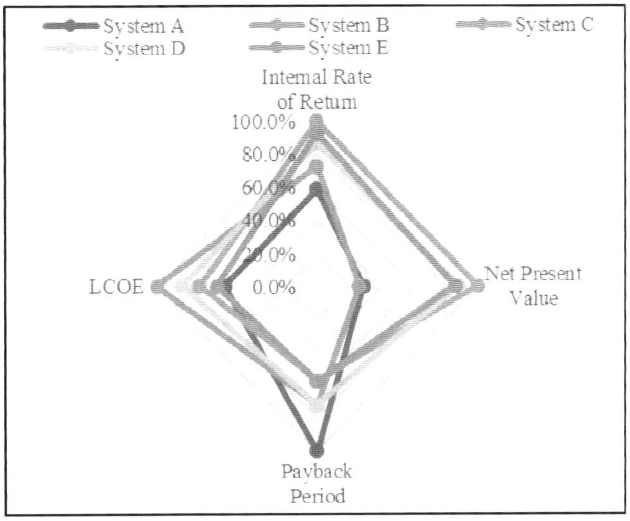


**Fig. 4.** Economic Comparison of the Defined Systems

The better economic parameters, with the lowest payback period and the highest IRR and NPV makes system C as the most economically attractive among all the hybrid technologies. Incidentally, this system also represents the equal share of capacities of the component technologies of the hybrid system. This additionally helps to balance the techno-economic drawbacks of the individual systems, resulting in higher economic performance.

**Table 4** Economic performance of System B and C with a variation of Cost of Biomass Feedstock

| Cost of Biomass Feedstock | 40 $/tonne | | 50 $/tonne | |
|---|---|---|---|---|
| | System B | System C | System B | System C |
| IRR (%) | 18.4 | 25.4 | 9.7 | 22.8 |
| Net Present Value (mn $) | 0.62 | 2.38 | -0.056 | 1.87 |
| Payback Period | 5 | 4 | 6 | 5 |
| LCOE (c$/KWh) | 14.34 | 10.50 | 16.58 | 17.70 |

The impact of the variation of the cost of biomass feedstock was studied and the results are shown in Table 4. The impact of the cost of biomass feedstock on a stand-alone biomass power system is much higher than the hybrid system, namely System C increasing the chances of economic failure of a standalone system. The result also shows that the hybrid system is much more resilient to individual system performance variations, as well as against unwanted externalities. Hence, an equal share of capacities in the hybrid system, as seen, would be able to optimize the induvial benefits and drawbacks and present an optimal economic solution for a small-scale grid-connected hybrid power generation unit.

## 5.3 Operation Strategies

Unlike individual systems, the output from the hybrid system can be effectively controlled to achieve better technical and economic benefits. Between the three hybrid systems studied, the hybrid system with an equal capacity of solar PV and biomass gasification unit has the potential for the highest flexibility of operations. By modulating the biomass power unit, and with complementing capacity factors, the hybrid 2 MW system, with an equal share of capacity between solar PV and biomass gasification, can be set to produce a stable quantity of electricity, even without storage systems like batteries. Due to the robustness of design of downdraft gasifiers and high part load efficiency of operations of IC Engines, such options are feasible. Hybrid operations of PV during the day and the biomass during the night would, therefore, be an interesting strategy that can allow the integration of such systems to an off-grid electricity network in future.

### 5.4 Carbon dioxide Mitigation Potential

The potential of decreasing carbon footprints of operations of an SME by renewable energy generation systems was thus analysed. A 3 MW Solar PV power plant is unable to supply the necessary energy demand. Even though the energy generated, amounting to 4.6 GWh is carbon neutral, the remaining energy is consumed at the rate of grid emission factor. Hence the overall system emission saving potential of System A was calculated to be 3536.9 tonnes per year, as shown in Fig. 5, with regards to the specific emissions of the Sri Lankan grid at 768.9 grams per KWh. A net emission reduction of 82% results thus.

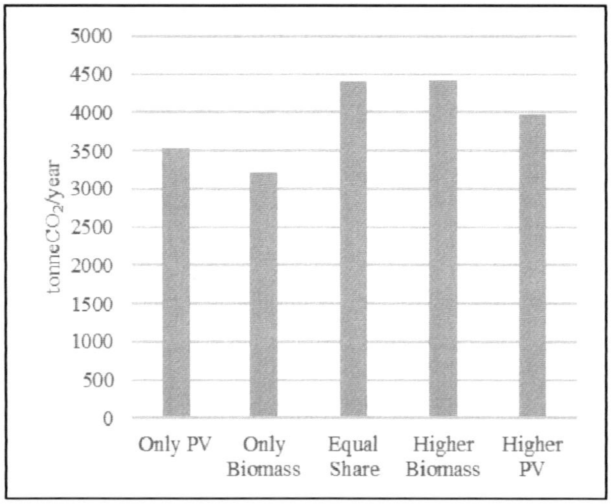


**Fig. 5** Emission saving Potentials of the compared systems

Net annual emission of the stand-alone 1 MW biomass gasification power plant fed only with energy crops or agricultural wastes was estimated at 800 tonnes CO2 based on (1). However, the presence of carbon-neutral waste paper, amounting to 20% of energy produced decreases the total emissions further. Therefore, an emission reduction of more than 80% using dedicated energy crops and agricultural wastes, in the form of coconut shells, along with waste paper was calculated to be feasible. Higher energy yields would result in higher absolute emission reduction, as is evident from results between system C and D. However, the increase of biomass power capacity within the hybrid system undermines the increasing energy generation, due to higher specific emissions of the biomass-based system with regards to solar.

### 5.5 Social and Other Impacts

The land requirement and other social and environmental impacts were assessed in brief. A lower land requirement for biomass power generation, 2 ha/MW would be advantageous against stand-alone solar systems to produce the same energy. However, with a hybrid system, land requirements for administrative purposes and electric switchgear can be shared among the technologies, decreasing the overall need for land further.

Overall environment impact analyses of the different systems were studied as well. Even though the impact on biotic and abiotic resources was not severe for all the systems, social impact and alteration of the local living patterns might occur. Particularly, inadequate system design of a biomass power system leading to liquid and gaseous effluent emissions might lead to serious social issues, severely affecting the long-term smooth operation of the biomass power, and hence putting the overall project at risk.

## 6 Conclusions

An analysis of the different small-scale renewable power generation systems was performed specifically to a case study in Sri Lanka. Solar PV and biomass gasification were obtained as the two most suitable technologies for small-scale applications. Grid integration would pose a significant challenge in the future, if not in many situations at present, especially with fluctuating renewables. Among the different hybrid system combinations, a 2 MW hybrid power generation with an equal capacity between solar PV and biomass gasification system resulted in the most suitable option for a small-scale grid-connected hybrid renewable power plant. A superior economic performance, besides robustness of the selected technologies, would further lead to easier installation and operation of such systems by SMEs. A minimum of 80% reduction potential of carbon dioxide emissions for the operations of SMEs was obtained through the hybrid power generation systems. Lower land requirements would be an added advantage towards sustainability of such small-scale power generation units.

## Acknowledgements

This work has been carried out as an extension of the industry involved academic/professional skill development project "Grand Challenge Project in Sustainable Energy System Development" performed by authors.

The financial, logistical and professional support provided by InnoEnergy (Sweden), and Wijeya Newspapers Ltd. (Sri Lanka) are gratefully acknowledged.

The authors would acknowledge the professional contributions by MSc students; Sachintha Rathnayake, Akila Fernando, Lalitha Srilal, Lathika Attanayake, Mihirani Kethumalika and Sachini Wimalasiri who were part of the team worked on the project "Grand Challenge Project in Sustainable Energy Development" focusing on finding sustainable energy solutions for the Sri Lankan company Wijeya Newspapers Ltd.

The valuable mentorship of the SELECT Grand Challenge projects generally provided by Prof.Torsten Fransson of KTH is specially acknowledged.

## References

[1] The Global Commission on the Economy and Climate, "Better Growth, Better Climate: The New Climate Economy Report: Energy," The Global Commission on the Economy and Climate, 2014.

[2] International Energy Agency, "World Energy Outlook Special Report: Energy and Climate Change," International Energy Agency, Paris, 2015.

[3] International Energy Agency (IEA), "Key World Energy Statistics," International Energy Agency, 2017.

[4] Organisation for Economic Co-operation and Development (OECD), "Enhancing the

Contributions of SMEs in a Global and Digitalised Economy," Organisation for Economic Co-operation and Development (OECD), Paris, 2017.

[5] J. Driesen and R. Belmans, "Distributed Generation: Challenges and Possible Solutions," in *Power Engineering Society General Meeting, IEEE*, Montreal, 2006.

[6] Sri Lanka Sustainable Energy Authority, "Sri Lanka Energy Balance," 2014. [Online]. Available: http://www.info.energy.gov.lk/. [Accessed 22 03 2017].

[7] E. Bocci , M. Sisinni , M. Moneti, L. Vecchione , A. Di Carlo and M. Villarini , "State of the art of small-scale biomass gasification power systems: a review of the different typologies," in *68th Conference of the Italian Thermal Machines Engineering Association, ATI2013*, 2013.

[8] Board of Investment of Sri Lanka, "Investment Guide," Research and Policy Advocacy Department of Sri Lanka, 2017.

[9] P. M. de Almeida, P. G. Babosa, C. A. Duque and P. F. Ribeiro, "Grid connection considerations for the integration of PV and wind sources," in *IEEE Xplore*, Bucharest, 2014.

[10] D. Elliot, M. Schwartz , G. Scott, S. Haymes , D. Heimiller and R. George, "Wind Energy Resource Atlas of Sri Lanka and the Maldives," National Renewable Energy Laboratory, 2003.

[11] Solar Energy Industries Association, "Solar Industry Data," Solar Energy Industries Association, 2016. [Online]. Available: https://www.seia.org/solar-industry-data. [Accessed 15 10 2017].

[12] Ministry of New and Renewable Energy, Government of India, "Biomass Gasification," Ministry of New and Renewable Energy , 2017. [Online]. Available: http://mnre.gov.in/schemes/grid-connected/biomass-gasification/. [Accessed 29 05 2017].

[13] Food and Agriculture Organization of the United Nations, "Coconut based farming systems and economics of CBFS," in *Consultation on Coconut Sector Development in Asia and the Pacific Region*, Bangkok, 2013.

[14] K. Janardhan , T. Srivastava , G. Satpath and K. Sudhakar, "Hybrid Solar PV And Biomass System For Rural Electrification," *International Journal of ChemTech Research,* **vol. 5, no. 2**, pp. 802-810, 2013.

[15] Obayashi Corporation, EX Research Institute Ltd, "10MW – Scale Biomass based Power Generation; JCM Feasibility Study (FS) 2014 – Final Report," The Joint Crediting Mechanism, 2014.

[16] Sri Lanka Sustainable Energy Authority, "Home," 2016. [Online]. Available: http://www.energy.gov.lk/.

[17] A. S. V. AG, "A Life Cycle Assessment of the production of a daily newspaper and a weekly magazine," AXEL SPRINGER VERLAG AG, Zürich, 1998.

[18] J. Sweeting, Project Cost Estimating: Principles and Practice, Rugby, Warwickshire: Institution of Chemical Engineers, 1997.

# Demonstration of a microgrid manager in a distribution grid with high PV penetration and Storage: La Graciosa case study

*Guillermo* Domínguez-López[1,*], *Pol* Paradell-Solà[1,*], *José Luis* Domínguez-García[1,*], *Jacob* Rodríguez-Rivero[2], and *Jorge* Sánchez-Cifuentes[2]

[1]Catalonia Institute for Energy Research (IREC), 08930 Sant Adrià de Besòs, Spain
[2]Endesa Distribución, Spain

**Abstract.** Future power systems with high penetration of distributed energy resources (DER) and information and communication technology (ICT) will allow increasing the controllability and observability of the grid. However, new challenges for low voltage distribution networks has arisen where the increasing DER is mainly photovoltaic (PV) installed on the roofs of buildings. This type of generation is variable and generates disturbances in the network. Islands are ideal to experience and install new technologies since are weak grids and the penetration of PV will lead to bidirectional power flows at distribution level, thus a smart control will be required. This article aims to present and innovative tool for the distribution system operators (DSOs) to monitor and manage grid operation. In near future, both DER and client's flexibility will increase. New monitoring and control tools for DSOs are essential. Furthermore, the grid operation optimization is needed to accommodate the expected DER installations and the active participation of consumers' flexibility services in markets ensuring system efficiency. This tool is tested on "La Graciosa" grid, one of the Canary Islands, to show its benefit. The grid studied includes a hybrid energy storage system (HESS), PV generators, controllable/uncontrollable loads and equipment to monitor power flows.

## 1 Introduction

Distributed energy resources (DER) are expected to take the lead of future power systems which have to be prepared for the increasing penetration of distributed generation (DG) at distribution level of the grid. A group of DER and loads in the distribution system is a microgrid (MG) [1-5]. A lot of methods and power management strategies are being developed [5-10] including DG, storage, controllable loads and other elements. However, these studies are focused on develop optimization methods for MG in simulation. There is a need for new monitoring and control tools for distribution system operators (DSOs) and land these tools from concepts to real world.

* Corresponding authors : gdominguez@irec.cat, pparadell@irec.cat, jldominguez@irec.cat

This paper presents a microgrid manager (MGM) tested in a pilot experiment done on "la Graciosa" island, which is located in the Canary Islands. It is worth noting that "La Graciosa" is the perfect network to validate and demonstrate the advantages of microgrids. The MGM consists on a microgrid optimizer and a supervisory control and data acquisition (SCADA) application. The performance and benefits of the MGM deployed are mentioned and also a brief explanation of lessons learned from the pilot that need to be considered when moving from simulations to real.

Part of this study has been carried out and validated on Project "GRACIOSA" which main partners are "Endesa Distribución", "Endesa Energía", "Cuadros Eléctricos Nazarenos" (CEN) and "Instituto Tecnológico de Canarias" (ITC) together with associate partners as IREC, Comillas, FIDETIA, ULL and AICIA.

## 2 Microgrid optimizer

The microgrid optimizer manages all the flexible assets existing on the distribution grid to make the network more reliable. It runs each 15 minutes and makes an optimization of 24 hours in steps of 15 minutes. A scheme of how data is flowing from the sensors to the set points send is shown in Fig 1. The devices considered in the system are explained in the next section with the SCADA and test scenario.

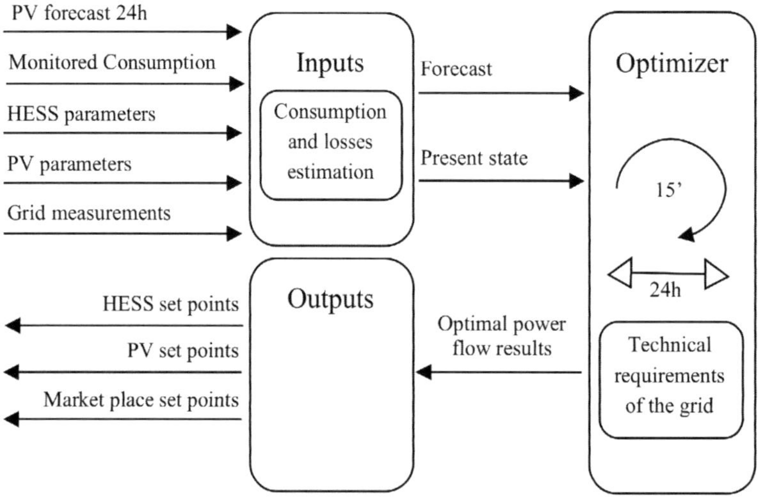


**Fig. 1.** Optimization method from sensor to set point.

The sensors provide data from the devices connected to the microgrid and grid measurements done along the system. Also, a PV forecast is used to estimate the consumption and losses of the microgrid. With historical data, a forecast for the consumption is also done and these inputs are served to the optimizer to run the optimal power flow. The results for the 24 hour prediction optimal power flow of the microgrid are gathered to send the appropriate set points to the devices at the correct time.

The optimization uses a three-phase balanced AC model of power flow with polar variables. It solves a multi-objective optimization maximizing the PV generation and minimizing the power losses of the distribution grid. The equations can be seen at:

$$\max f(PGEN) = \sum_{t\in[1,nt]}\left(\sum_{i\in[1,nPV]} PV_{i,t} - \sum_{(n,k)\in L\cup(i,i)} Ploss_{n,k,t}\right) \quad (1)$$

$$Ploss_{n,k,t} = G_{n,k}\left(V_n \sin\delta_{n,t} - V_k \sin\delta_{n,t}\right)^2 + B_{n,k}\left(V_n \sin\delta_{n,t} - V_k \sin\delta_{n,t}\right)^2 \quad (2)$$

Also to accomplish the purpose of the microgrid optimizer, the model has physical constraints as voltage limits of the network, power limits of each branch and the state of charge limits for the batteries.

The language used to manage this program is Julia with some complements. Julia is a high-level dynamic programming language designed to address the needs of high-performance numerical analysis and computational science. Also, the program uses the Bonmin solver, installed on Julia, to resolve the non-lineal equations of the model.

## 3 SCADA and test scenario

In this section, the test scenario and the SCADA are presented. Two lines with high PV penetration of the whole distribution grid of the "La Graciosa" are in the scope of the pilot and they can be seen together with the equipment installed in the grid in Fig. 2 which is one of the screens of the SCADA developed in an industrial tool (Wonderware by Schneider Electric). The system consists of 6 power analysers, a Hybrid Energy Storage System (HESS), 2 PV installations, 2 cameras for PV forecast and, finally, 3 self-consumption (PV with storage) installations have been deployed in clients to provide flexibility.

All devices have been integrated in the SCADA so the application can send set points from the optimizer and store data in real time, as it can be seen in Fig. 3. The communications from the SCADA server to the equipment deployed is based on Modbus TCP and connected by 3G to internet.

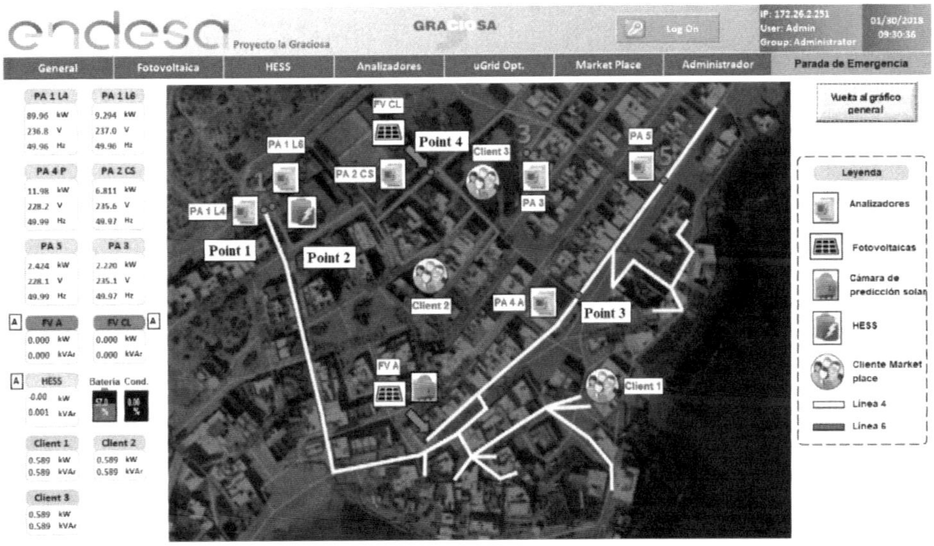


**Fig. 2.** Lines involved and equipment installed in "la Graciosa" island.

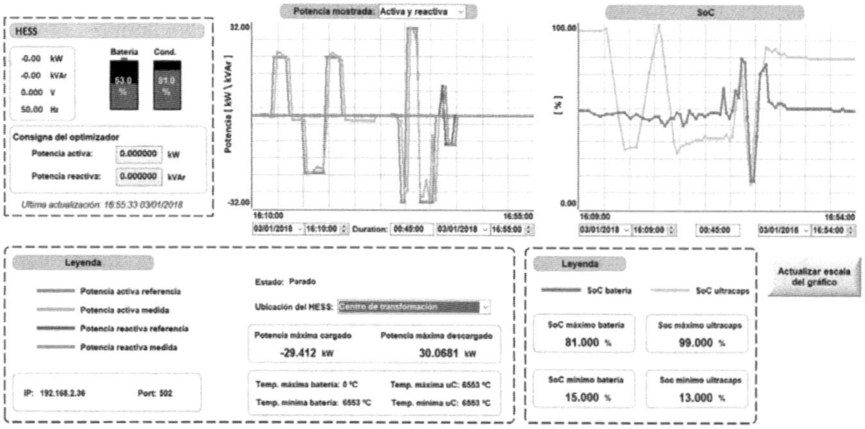


**Fig. 3.** Real-time monitoring of HESS performance during the test.

## 4 Tool capabilities and main test results

The equipment installed and the tool deployed increase the grid functionalities and upgrade the system into a smart microgrid. The SCADA enables to manage the system locally and remotely while it gathers information from the equipment in the grid to real-time monitor it. Therefore, the DSO is able to detect grid events at distribution level and study data stored to enhance power supply security and reliability. To illustrate this point, a discharge of the grid during the test is shown in Fig. 4 where the voltage and the power flowing in the line droop to zero and autonomously the tool reconnects all the equipment when the power supply is restored.

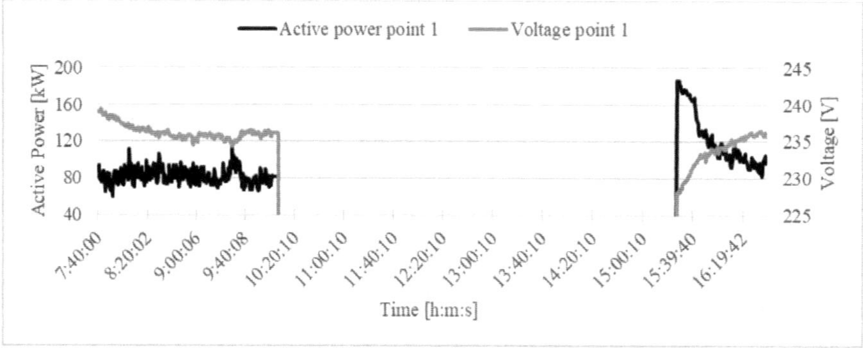


**Fig. 4.** Real-time monitoring of a discharge and reconnection during the test.

Moreover, the SCADA offers the possibility of remotely operate DER and HESS to change their low level control strategy and turn on, turn off or reset the device if a malfunction appears during real-time management. This new asset avoids to send a technical operator by the DSO for minor operations and offers an increase of the system controllability.

Moving to the main test results of the optimizer, the first indicator that is calculated is the accuracy of the optimizer prediction. The accuracy is defined as equal to the error between the prediction expected by the optimizer and the real power measured by the power analysers. These errors calculated at some measurement points provide a general overview of the accuracy of the PV and demand forecasts and, in fact, the performance of the optimizer and set points sent to the smartgrid active actors.

As usual when simulation analysis reach a pilot, the test came across real world unexpected problems which had to be solved as they showed up. The main problems faced during the test were communication problems between the SCADA server and the rest of the equipment due to weak 3G internet connection in the island. These problems had interaction with the PV and demand calculation, but as the accuracy is intended to be calculated considering normal operation with no communication problems, the affected values by these problems are underestimated. The accuracy is calculated for 4 points of measure that can be seen in Fig. 2. The function which calculates the accuracy is defined by equation 3 and its values are presented in Table 1.

$$\text{Error}(t) = [Value\_measured(t) - Value\_prediction(t)\ ]/(Nominal\ Line\ Power)^{\dagger} \quad (3)$$

**Table 1.** Maximum and average error calculated for the most relevant measurement points.

| | Point 1 (Line 4) | Point 2 (Line 6) | Point 3 (Line 4) | Point 4 (Line 6) |
|---|---|---|---|---|
| Maximum % Error | ±22 % | ±16 % | ±9 % | ±10 % |
| Average % Error‡ | 7% | 4% | 2% | 4% |

The average error is lower than 7% for all the measurement points, so the accuracy of the microgrid optimizer is high given the number of measurement points for the two lines and the uncertainty of the system. The power analysers have an own confidence interval in its measurements, usually ±0,5 %, so the average error of the optimizer is potentially lower if this tolerance is removed. Furthermore, it can be seen in Fig. 5 how the error is mainly greater during the PV production hours and the maximum error found from the whole demonstrator is ±22 %. This fact is related with several variables that can influence, but it is mainly a sign of the accuracy of the demand and PV forecast which is highly erratic.

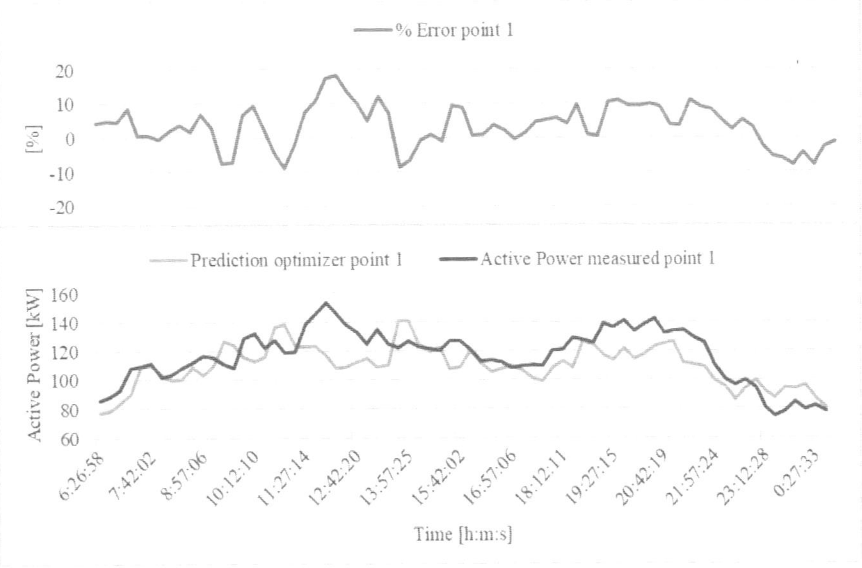


**Fig. 5.** Active power prediction, real value and error in time domain for point 1.

† Nominal Line Power: Nominal power value of each line depending on point of measure.
‡ Average error: Calculated through the root mean square error.

In addition, another study is done to check if the clients of the Market Place are able to follow the set points calculated by the microgrid optimizer. The algorithm calculates the maximum power consumption of the clients to reach the technical requirements of the grid so, theoretically, the Clients should not exceed this consumption. The consumption and the maximum power consumption expected of two Clients are shown in Fig. 6.

The three Clients follow the maximum power consumption requirement from 87% to 91% of the test time. Thus, the consumption expected is exceeded between 9% and 13% of the total time and this fact affects on the microgrid optimizer accuracy. In addition, the maximum error between the measured consumption and the maximum expected consumption calculated by the optimizer is presented in Table 2. The function to calculate the error is given by equation 4.

$$\text{Error}(t) = [Cons\_meas(t) - Max\_cons\_predic(t)]/(Contracted\ Power)^{\S} \quad (4)$$

**Table 2.** Maximum error calculated for the clients of the market place.

| | Client 1 | Client 2 | Client 3 |
|---|---|---|---|
| Maximum % Error | 1,8 % | 69 % | 54 % |

This error can be up to 69%, so it is clearly a source of deviations in the microgrid optimizer and, as a consequence, the set points sent to the PV inverters and to the HESS do not reach the optimum point of operation of the system.

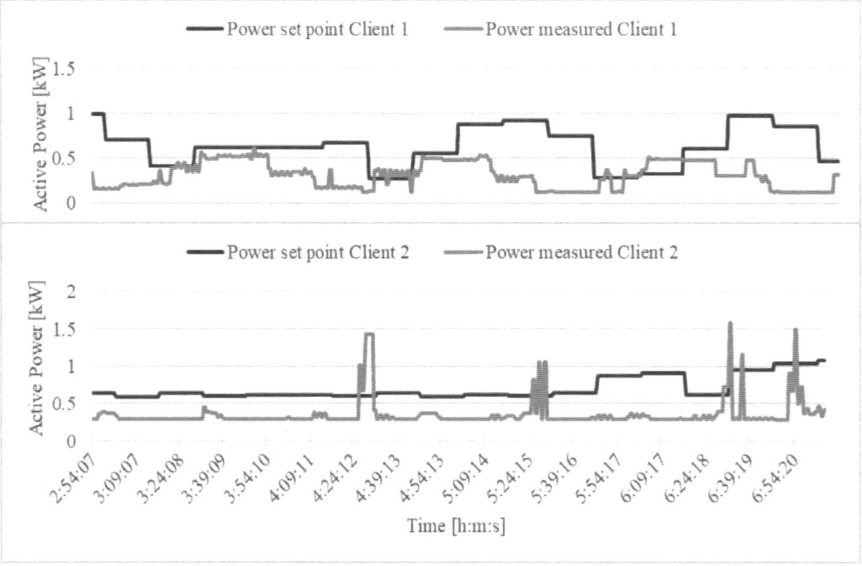


**Fig. 6.** Power consumption and maximum consumption expected for clients 1 and 2 in time domain.

## 5 Conclusion and next steps

This paper proposed an optimization strategy for microgrids with DER and HESS together with a centralized communication structure based on a SCADA application. Both together, optimization and SCADA, have been tested in a real pilot on "la Graciosa" island. Finally, a SCADA is found to be a great implementation to upgrade conventional distribution grids into

§ Contracted Power: Considered as each Client Contracted Power.

smartgrids. The accuracy of the predictions of the optimizer and its performance is acceptable, but highly dependent on DG and demand forecast accuracy. Real-time low level controls are needed to compensate the error given by a non-accurate forecast or communication problems during normal operation. In a future local market where clients will be benefited for offering flexibility to the system, they should be penalized when they do not accomplish the requirements.

There are mainly three actions that could potentially upgrade the automatization process and better performance of the optimizer. A first action would be to upgrade the communications structure switching 3G connection into a wifi to connect all the devices together Also, a suitable protocol for internet of things such as LoRa would increase the reliability of the remote connection and quality of the data. The optimization is highly dependent on the forecast for both solar and consumption, so a second action would be to improve the knowledge of the distribution grid, the forecasting and deployment of more points of measurement along the lines. Finally, a real time module could be implemented to balance the power set points calculated to adjust the forecast with the real powers flowing in the microgrid.

Future work is focused on improve the system with three phase imbalance conditions as the low voltage microgrid studied has several loads single-phase connected. Currently the system is studied under normal conditions of the microgrid, but the microgrid needs to be improved and test under emergency state. Also, a next step would be integrate more renewable generation and test the system islanded from the main grid. Nevertheless, an implementation of a strategy like Machine Learning to boost the system forecast for PV and consumption would be a next step into grid state estimation minimizing the power analysers deployed.

## Acknowledgments

This research has been carried out by Project "GRACIOSA" funded by the programme INTERCONNECTA (FEDER-CDTI funds).

## References

1. N. D. Hatziargyriou and A. P. S. Meliopoulos, "Distributed energy sources: Technical challenges". *IEEE Power Eng. Soc. Winter Meeting*, **2**, 1017–1022, (2002).
2. R. H. Lasseter and P. Piagi, "Microgrid: A conceptual solution". *IEEE Power Electron.*, **6**, 4285–4290, (2004).
3. H. Jiyaji, J. Chuanwen and X. Rong, "A review on distributed energy resources and MicroGrid". *Renew. Sust. Energ. Rev.*, **12**, 2472–2483, (2007).
4. S. M. Nosratabadi, R. Hooshmand, and E. Gholipour, "A comprehensive review on microgrid and virtual power plant concepts employed for distributed energy resources scheduling in power systems". *Renew. Sust. Energ. Rev.*, **67**, 341-363, (2017).
5. F. Katiraei and M.R. Iravani, "Power Management Strategies for a Microgrid with Multiple Distributed Generation Units". *IEEE T. Power Syst.*, **21**, 1821-1831, (2016).
6. A. Maulik and D. Das, "Power Flow and Economic Dispatch of a Microgrid in Presence of Feeder Flow Control Distributed Generators". *2017 6$^{th}$ International Conference on Computer Applications In Electrical Engineering-Recent Advances (CERA)*, 187-192, (2017).
7. J. Qiu, J. Zhao, Y. Zheng, Z. Dong, Z. Y. Dong, "Optimal allocation of BESS and MT in a microgrid". *IET Gener. Transm. Dis.*, **12**, 1988-1997, (2018).

8. A. Maulik and D. Das, "Optimal operation of microgrid using four different optimization techniques". *Sustain. Energy Techn. Assess.*, **21**, 100-120, (2017).
9. M. Lei, Z. Yang, Y. Wang, H. Xu, L. Meng, J. C. Vasquez and J. M. Guerrero, "Design of Energy Storage Control Strategy to Improve the PV System Power Quality". *IEEE Ind. Elec.*, 2022-2027, (2016).
10. Y. Zhu, F. Zhuo, H. Shi, "Power Management Strategy Research for a Photovoltaic-Hybrid Energy Storage System". *2013 IEEE ECCE Asia Downunder*, 842-848, (2013).

# Evaluation of a hybrid system to improve the electrical efficiency in photovoltaic panels

*Juan C* Valencia[1,*], *Mauricio* Diaz[1], and *Carlos* Giraldo[1]

[1]Pontificia Universidad Javeriana Cali, College of Engineering, Cali, Colombia.

**Abstract.** This project shows a photovoltaic hybrid system (PHS) that follows the sun using electronic and mechanic devices and reduces the solar panel's temperature with water in order to increase the energy obtained. The system was designed and implemented with a 1-axis solar tracker system and a water cooling system. PHS allows solar radiation beams to fall almost perpendicularly and decreases the temperature of the solar panel. Finally, its performance is compared against a traditional fixed photovoltaic system (FPS) oriented at 15° facing south. Experiments were made during only seven days in a year between December 1st and December 19th of 2017 in Guadalajara de Buga, Valle del Cauca, Colombia. Tests consisted of using the same electric loads at 20%, 40% and so on until 100% of the total power capacity of the solar panels. Results points out that electrical efficiency of the PHS increased 5.03% compared to the FPS on the same environmental conditions. Finally, PHS shows daily additional average power between 37% and 40% when it is loaded with electric loads at 80% and 100%.

## 1 Introduction

Big part, almost 79% of world energy production and consumption is derived from fossil fuels like oil, coal and gas; 19.3% is derived from renewable energy and a minor part with only 2.3% is derived from nuclear energy [1]. The problem with this consuming and generating scheme of energy is the negative effect on the earth climate and the contamination that it generates which affects vegetable, animal and human life. Thus, it is necessary to look for new laws and strategies to deploy the use of renewable energies.

A bunch of governments and people around the world have taken consciousness about this problem and they are redirecting its actions toward the use of renewable energies. On these days, exists different technologies to produces energy from natural resources as the sun, water, wind, decomposition and thermal heat. Among all the renewable technologies, the most popular is the photovoltaic solar energy which had a significant grow of 47% in 2016. Instead, wind and hydropower had a grow of 34% and 15.5% according to the information provided by [1].

* Corresponding author: jcvalencia22@hotmail.com

Despite this preference on photovoltaic solar energy there is a couple of issues that affects negatively its energy production: In the process of producing energy, solar panels are affected by the not perpendicularity incidence of solar radiation beams due to the relation movement of earth and sun, and the rise temperature effect gained through the exposition of solar radiation during the day.

Taking account, the aforementioned problems, a hybrid system is designed, implemented and characterized to follow the sun using a mechanical 1 axis solar tracker and integrated with a water cooling system to refrigerate the solar panel surface. The main goal is to know how much energy can produce our system against a traditional photovoltaic system. The figure 1 shows both systems.

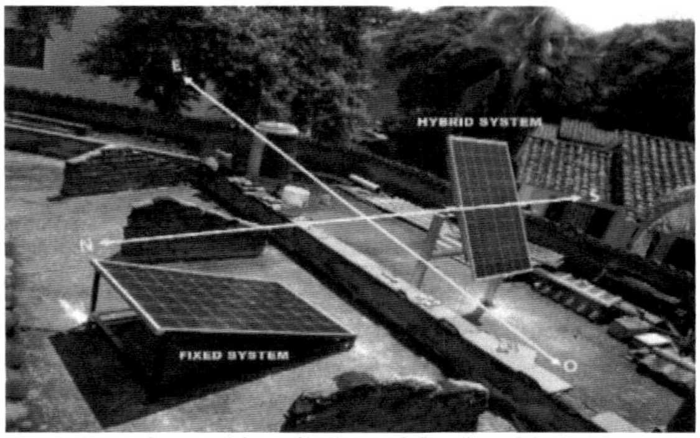


**Fig. 1.** Location and installation of fixed and hybrid system.

## 2 Implementation of Hybrid System

The hybrid system aims to maximize the energy production, minimizing three phenomena that affect the uptake of energy: the non-perpendicular incidence of the solar radiation beams on the solar panel, for which a 1-axis solar tracker was constructed to follow the sun during the day; the negative influence of the temperature with respect to the power generated by the solar panel and the losses by reflection, and due to these last two factors a water cooling system was implemented, in order to decrease the operating temperature of the panel and also reduce the losses by reflection, this is due to the refractive index of water that is 1.3, intermediate medium between the air which has a refractive index of 1 and the panel glass with an index of approximately 1.5.

Thus, the hybrid system is integrated by three subsystems: the tracking system, the cooling system and the data acquisition system. See Figure 2.

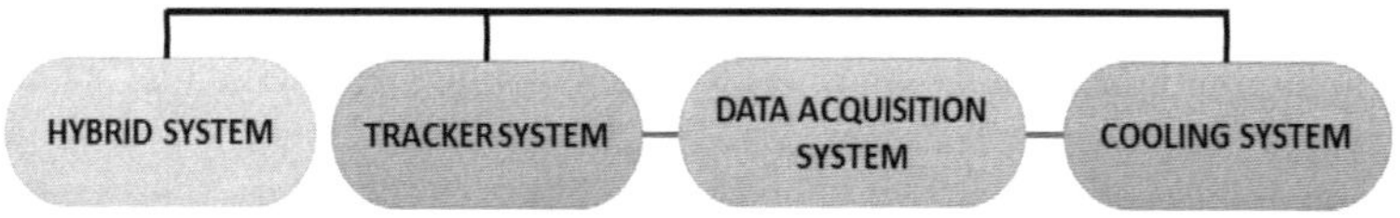


**Fig. 2.** Hybrid system is integrated by a solar tracker system, a water cooling system and a data acquisition system.

### 2.1 Tracker System

The tracking system implemented is one degree of freedom, it is responsible for moving the panel through the day to increase the incidence of solar radiation beams on it. To carry out the movement, the tracking system is composed of a mechanical system, which supports the solar panel, a DC electric motor, a DC power supply and finally the electronic system (power, sensors and control subsystems).

### 2.2 Cooling System

The cooling system is responsible for cooling the solar panel front surface by generating a film of water which was pumped by two DC water pumps. These pumps were placed in a clay container that stored the water. In addition, the system has a closed circuit of water circulation that allows water to circulate from the storage container, passes through the panel and finally returns to the container. The water recovery system was achieved by putting drains to each end-side of the panel.

### 2.3 Data Acquisition System

The physical variables measured for the traditional fixed system are the temperature in the rear center part of the panel, voltage and current consumed by the electric load. For hybrid system they are: temperature in the rear central part of the panel, water cooling temperature, water drain temperature, environment temperature, solar radiation, the voltage and current that is consumed by the electric load, voltage and current consumed by the DC motor and the angular position of the system.

## 3 Tests

In Colombia, due to the location next to Ecuador, the hours on which it is possible to have solar radiation are approximately from 6:00 AM to 18:00 PM. According with the information provided by UPME [6], the daily average range of solar radiation energy on December is from 4.0 to 4.5 kWh/m2. Our measured data says that the average amount of solar energy is 4.643 kWh/m2. In addition, in the same way as Ramkumar et.al [5], which carried out the experiment in five days and acquired data in the morning from 9:00 AM to 4:00 PM in the evening. Also, in [3] the measurements were performed during the month of February and recorded data from 9:00 AM to 4:00 PM. Therefore, due to the arguments mentioned before, the behavior of both systems were analyzed in seven days from 8 am. to 4 pm. GMT-5. The tests consisted in each day connecting an electric load of the same value to each system, whose objective was to analyze the both systems' performance measuring nominal power (Watts). Initially, an electric load demanded approximately 20% of the nominal power of each system that is 51W of 255W, the second day, an electric load of 40% demanded 102W of 255W and so on until reaching a load that demanded 100% of the nominal power of the system. All samples were carried out under the same environmental conditions and the work point of both solar panels were under the same electric load value. This experiment was done in order to analyze the behavior of both systems' generated power when the consumption, given the load, was close to the limit of both solar panels' capacity. This way, as the electric load demanded more power, it was expected that the hybrid system would be able to deliver more energy than the traditional fixed system. To analyze the performance, the power delivered by each system was sampled every 5 minutes and the data was sent via WiFi and stored in a spreadsheet on Google Drive.

## 4 Results

In this section the results are shown in terms of solar radiation measures, energy produced, electrical and thermal efficiency, fixed and hybrid system temperatures. There were 6,828 data collected in seven days of measurement, also, infrared pictures were taken that qualitatively shows the temperature behavior of both systems. Before showing the performance results, it is worth highlighting that the data acquired from the experiment contains different environmental conditions, from cloud days to sunny days, as well as the behavior of the shadows coming from cloud's journey. This kind of data let having a variety of days to support the experiment results under different weather conditions. The figures shown below support the description detailed before.

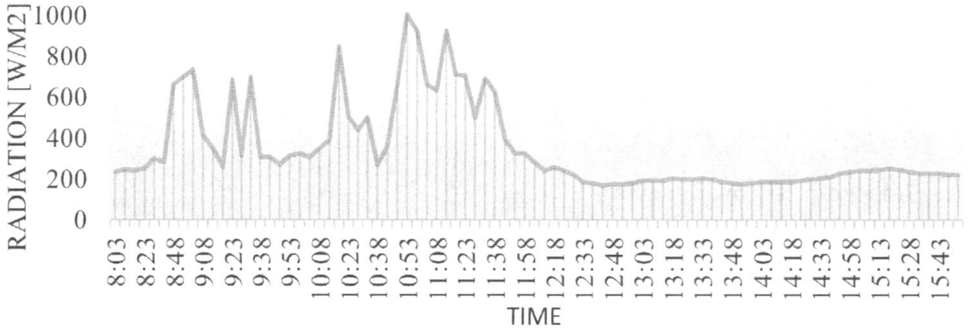


**Fig. 3.** Solar radiation behavior of December 1st of 2017.

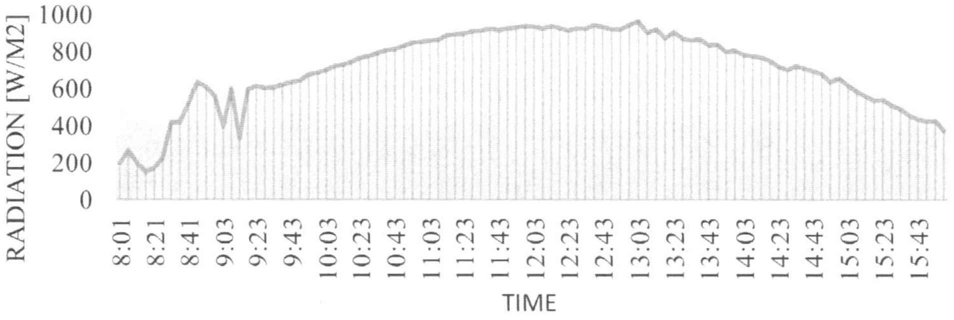


**Fig. 4.** Solar radiation behavior of December 19th of 2017.

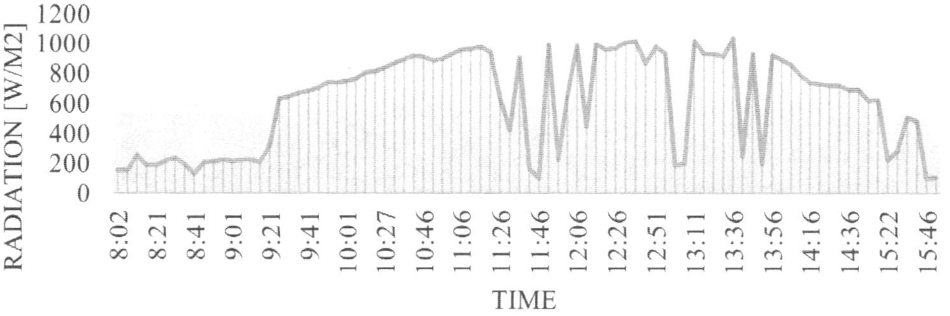


**Fig. 5.** Solar radiation behavior of December 16th of 2017.

Also, it should be mentioned that it was possible just to do a seven-day experiment, for the availability of the technical devices we had since they didn't belong to us and because of our economic resources that were too limited.

## 4.1 Electrical Results

Table 1 and Table 2 shows the data acquired about the average solar radiation, total energy produced and the average of electric efficiency for traditional fixed system and hybrid system during the day of experiment. Notice how energy produced by hybrid system is always greater than the fixed system (Table 1 and figure 6)

**Table 1.** Electric results of fixed system.

| Load | Average Radiation $[W/m^2]$ | Energy Fixed $[Wh/day]$ | Energy Hybrid $[Wh/day]$ | η Fixed | η Hybrid |
|---|---|---|---|---|---|
| 20% | 343.77 | 291.86 | 351.76 | 6.61% | 8.10% |
| 40% | 597.33 | 688.83 | 849.85 | 9.88% | 12.22% |
| 60% | 614.04 | 762.74 | 945.06 | 9.41% | 11.38% |
| 80% | 628.34 | 931.13 | 1250.11 | 11.23% | 15.47% |
| 80% | 628.34 | 1064.59 | 1455.29 | 11.93% | 16.96% |
| 100% | 521.05 | 703.30 | 989.20 | 8.22% | 11.92% |
| 100% | 661.91 | 1085.03 | 1467.28 | 12.55% | 17.33% |

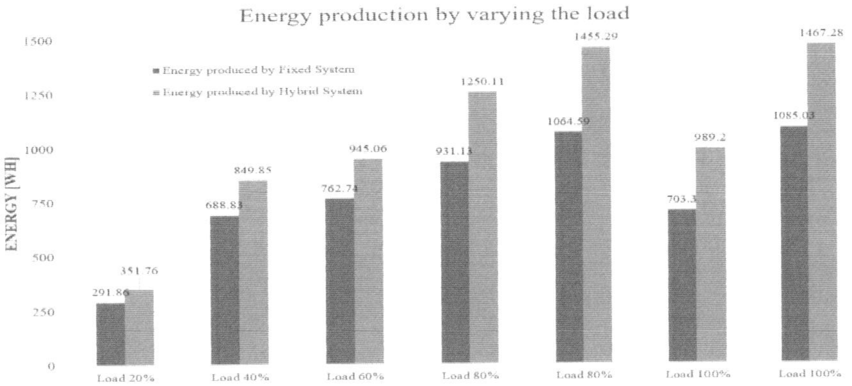


**Fig. 6.** Energy production of fixed and hybrid system by varying the load.

Analyzing the data about energy production in both systems, it is worth highlighting the power difference that the hybrid system could generated on the day, variable, which is called the additional percentage of power generated by the hybrid system over the fixed system. These numerical values are illustrated in Table 2.

**Table 2.** Additional percentage of power generated by the hybrid system over the fixed system.

| Load | Average Radiation $[W/m^2]$ | Additional Percentage of Power [%] |
|---|---|---|
| 20% | 343.77 | 22.70 |
| 40% | 597.33 | 23.38 |
| 60% | 614.04 | 23.90 |
| 80% | 628.34 | 37.15 |
| 80% | 628.34 | 40.48 |
| 100% | 521.05 | 40.85 |
| 100% | 661.91 | 39.52 |

From table 2, it is observed that best performances of the hybrid system occurred when the load is at 80% and 100% of the total energy capacity of solar panels. It is recalled that the measurements were made in equal environmental and operating conditions for both systems. Thus, figure 7 shows how are the production difference of the hybrid system during the day for all the tests. Here, best results occur at morning from 8:00 to 10:00, which can be explained for various reasons.

On those days close to winter solstice, a mismatch presented on the installation orientation together with "sun's movement" at the age the experiment was carried out and the connection of a high electric load favored the catchment of solar radiation beams during the morning for the hybrid system and thus its performance against the load was greater. As the hybrid system tracked the sun and as it had a high electric load that consumed more power, it could have more solar resources to deliver more energy than the fixed system. Instead, as fixed system does not track the sun, it is not well positioned from 8:00 to 10:00 to receive the solar radiation beams.

Finally, the water temperature of the cooling system at mornings was lower than at afternoons (the water acquired heat as the time passed), a condition that allowed to decrease the working temperature point of the hybrid system solar panel and hence have a better performance.

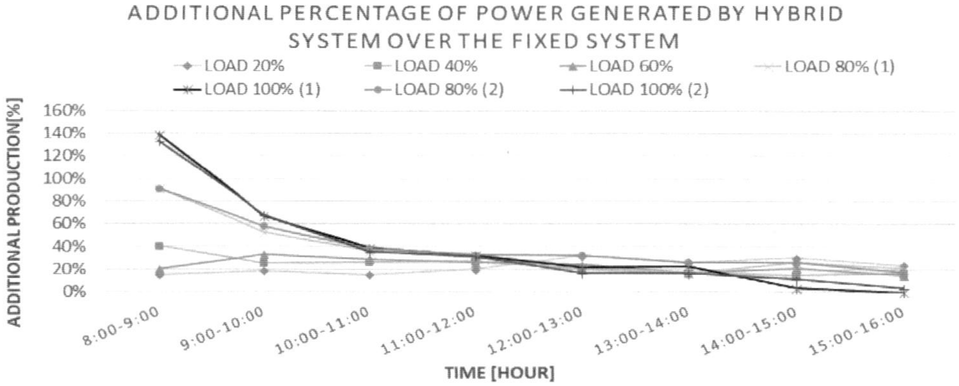


**Fig. 7.** Additional percentage of power generated by the hybrid system over the fixed system.

In addition, after 15:00, the figure 4 shows a decrement of additional production when the electric load connected to the hybrid system was 100% of the solar panel total capacity. This happened because both system was under rain conditions. For the rest of the days, the additional production was approximately 20%.

## 4.2 Thermal Results

Table 3 shows the average radiation for each experiment, the average of the ambient temperature, the temperature of the fixed system panel, the temperature of the hybrid system and its thermal efficiency.

**Table 3.** Thermal results of fixed and hybrid system.

| Load | Radiation $[W/m^2]$ | Environment Temperature [°C] | Hybrid Temperature[°C] | Fixed Temperature [°C] | ηth |
|---|---|---|---|---|---|
| 20% | 343.77 | 28.34 | 29.05 | 34.39 | 16.23% |
| 40% | 597.33 | 32.82 | 34.31 | 48.60 | 41.07% |
| 60% | 614.04 | 32.29 | 34.65 | 49.02 | 54.40% |
| 80% | 628.34 | 35.06 | 35.95 | 51.95 | 47.48% |
| 80% | 628.34 | 34.28 | 35.89 | 52.48 | 37.24% |
| 100% | 521.05 | 32.45 | 33.46 | 45.06 | 56.93% |
| 100% | 661.91 | 35.35 | 36.34 | 45.01 | 30.44% |

By observing the temperature value of the fixed system panel and the hybrid system panel it is possible to see that the panel temperature of the hybrid system is much lower than that of the fixed system and tends to be the ambient temperature. Regarding the relation of the thermal efficiency and the variation of the load, there is not a distinguished pattern, so the thermal efficiency depends, in this work, on the weather conditions.

Now, the influence of the temperature distribution in the panel of the fixed system and the hybrid system in figure 8 will be analyzed qualitatively, where figure 8.a shows a higher temperature distribution than the one of the hybrid system in 8.b, this last one shows the effect of water cooling. Finally, figure 8.c shows a thermal scale of the temperatures. Note: these pictures were taken with the FLIRONE camera with the recommendations specified in [2].

**Fig. 8a.** Fixed system **Fig. 8b.** Hybrid system **Fig. 8c.** Thermal scale

**Fig. 8.** Thermographic images of both systems and thermal scale with maximum and minimum values.

## 5 CONCLUSIONS

- The hybrid system discussed increased the energy production in 40% compare to the fixed system, working at 80% to 100% of load capacity.
- When a solar panel is directly connected to an electrical load of lower power consumption than the panel can deliver, it needs less radiation to satisfy the energy demand of the load.
- The electrical efficiency of the hybrid system over the fixed system could be up to 5.03%.
- Fixed mount photovoltaic installations oriented towards the south diminish their capture of solar radiation from the spring equinox to the autumn equinox in locations close to the equator, that is, from March to September, since this installation would not be to exposed to the direct radiation of the sun thanks to the apparent movement of this.

## References

[1] R. E. P. N. f. t. 2. Century, "http://www.ren21.net," 2017. [Online]. Available: http://www.ren21.net/wp-content/uploads/2017/06/17-8399_GSR_2017_Full_Report_0621_Opt.pdf.

[2] FLIR, "Thermal Imaging cameras: a fast and reliable tool for testing solar panels," [Online]. Available: http://www.flir.co.uk/instruments/building/display/?id=41872.

[3] H. M. Bahaidarah, S. Rehman and P. Gandhidasan, "Experimental evaluation of the performance of a photovoltaic panel with water cooling," [Online]. Available: https://scholar.google.com/scholar?hl=es&as_sdt=0%2C5&q=Experimental+Evaluation+of+the+Performance+of+a+Photovoltaic+Panel+with+Water+Cooling&btnG=.

[4] E. L. R. E. Network, "http://www.exa.unicen.edu.ar/," 2007. [Online]. Available: http://www.exa.unicen.edu.ar/catedras/tecnoambiente/eyp.pdf.

[5] M. K. C. R. A. R. R. Ramkumar, "Enhancing the performance of photovoltaic module using claypot evaporative cooling water," in *Energy Efficient Technologies for Sustainability (ICEETS)*, 2016, pp. 217-222.

[6] UPME, "https://biblioteca.minminas.gov.co/," [Online]. Available: https://biblioteca.minminas.gov.co/pdf/Atlas%20de%20radiaci%C3%B3n%20solar%20Colombia.pdf.

# The effect of building energy management systems on battery aging.

*Lluc* Canals Casals[1,*], *Lucía* Igualada[1], and *Cristina* Corchero[1]

[1]Catalonia Institute for Energy Research (IREC), Energy Systems Analytics Department, 08930 Sant Adrià de Besòs, Barcelona, Spain.

**Abstract.** Smart buildings are a key element to walk towards smart cities and grids. Nonetheless, there are several degrees of intelligence. A first step is to incorporate commercial self-consumption solutions in buildings so they can manage the energy from local renewable power generators. A second step is to substitute this commercial solutions with an optimized Energy Management System (EMS) to reduce the electricity bill at the end of the month. Further. This EMS may contribute to stabilize and improve the quality and emissions of the electricity grid by offering some energy flexibility to the electricity system in favour of decentralization. This study compares the battery aging between buildings that count with an EMS to optimize the electricity bill under three scenarios in contrast to those that have a simple self-consumption kit. Lithium ion battery lifespan is estimated by means of an electric equivalent battery circuit model that runs on Matlab and simulates its behaviour through time. Moreover, this study evaluates the distribution of the battery costs regarding its use, observing that batteries controlled by simple self-consumption kits have longer lifespan because they are underused, ending up in higher calendar aging costs than the ones that are controlled by EMS.

## 1 Introduction

Renewable energy sources can be integrated in any place in the electricity grid, from large-scale generators to residential buildings [1]. It is widely accepted that energy storage systems such as batteries are a key element to increase the integration rate of renewable power sources into the grid [2] and both contribute on the configuration of decentralized smart grids [3]. Thus, residential buildings would be the smaller element in decentralized smart grids of this kind, being the core of this study.

To manage the energy generation and demand of buildings, several devices can be used. Generally, buildings count on self-consumption kits that use a simple energy management program, such as the one presented in Fig 1, due to the lower investment costs. These systems store the excess of energy generated by the renewable power systems of the building in the energy storage system (generally a battery). If the battery is full, this excess is delivered to the electricity grid. While the battery has enough energy and the generation

* Corresponding author: lcanals@irec.cat

is lower than the building's energy demand, the battery will discharge. When the battery is completely discharged, it is the grid that should provide all the building energy demand.

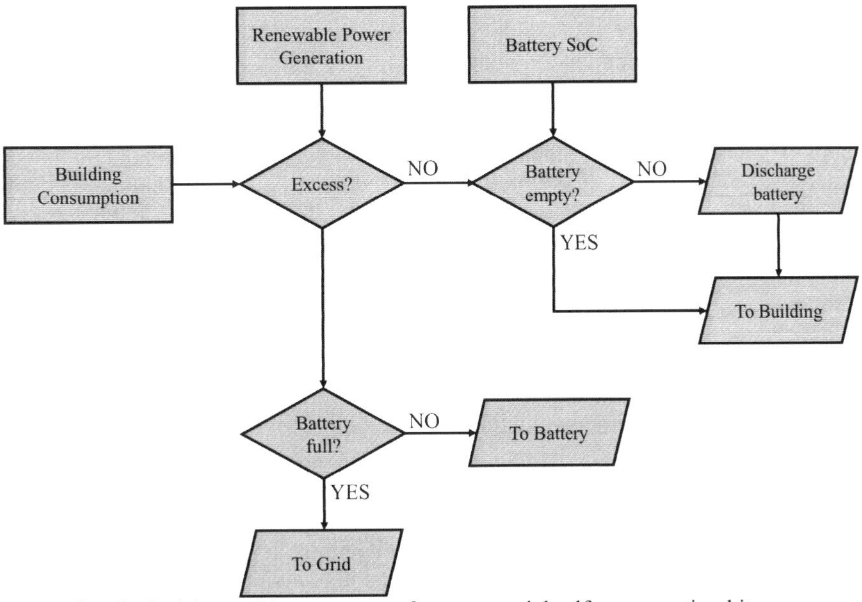


**Fig 1.** Example of a decision making program of a commercial self-consumption kit.

On the contrary, more complex Energy Management Systems (EMS) incorporate forecasting sources to predict weather conditions, electricity prices and building load in their decision making algorithms so they can reduce either the monthly electricity bill [4] or the greenhouse gas emissions [5].

These economically oriented EMS tend to use batteries much more often than the commercial kits do, which certainly has an impact on battery aging [6]. Although several V2G or intelligent electric vehicle charging systems do consider other aspects, such as battery lifespan enlargement, in the energy management algorithm [7], most of the algorithms implemented in households' EMS do not incorporate battery aging factors in their optimization calculations due to the complexity and particularity of ageing related to each battery chemistry.

Therefore, this study compares the effect that the different management strategies between commercial self-consumption kits and EMS under three different scenarios have on battery aging.

## 2 Methodology

This study uses monitored consumption data of the common areas (lighting, elevator, communications…) of a building during one year that is part of the GrowSmarter UE funded project. This building counts on 4 kWp photovoltaic generators on the rooftop, a self-consumption kit and a 6,5 kWh lithium ion battery that costs 4011€. The battery useful Depth of Discharge (DoD) is 90%, this means that the system can use only 90% of its nominal capacity. This is the base scenario of the study in which the battery is partially used, having an average use of 0,25 cycles per day.

A previous work [8] considered these daily generation and energy load of the building to simulate the behaviour of the EMS under three different scenarios (represented in Fig 2) and to compare the benefits it may provide against commercial systems:
- High (1): The battery is used whenever there is a positive balance in the electricity prices, doing an average of 2,4 cycles per day.
- Medium (2): The difference in price between the lower valley and the peak is higher than in the High scenario, thus, the smaller differences are not considered. In this case, the battery does over 1,5 cycles per day.
- Low (3): The battery is discharged only when the electricity peak cost is over 0,14€/kWh and it is charged at lower valleys. In this scenario, the battery does an average of 1 equivalent full cycle each day.

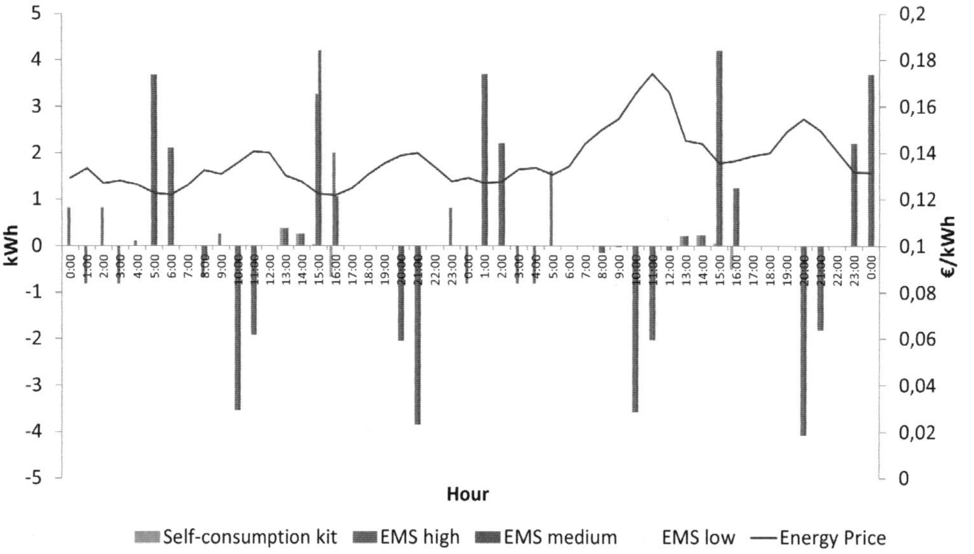


**Fig 2.** Representation of the battery use in all scenarios during 2 days (bars) and the electricity price (black line).

The EMS recalculates the optimal solution every 15 minutes considering inputs from the day-ahead tariff (economic), the forecasted solar energy generation, the expected battery SOC and adapting its results with the deviations occurred due to real energy consumption in the building and from what really occurs with the solar generation. Fig 2 represents the battery hourly energy exchange (bars) in all scenarios during two days extracted from the whole year monitoring and the corresponding energy price per hour (black line). Fig 2 shows how the EMS in scenario 1 (High) tries to use the battery far more often than in all the other cases. Moreover, Fig 2 shows how the commercial kit clearly underuses it and, moreover, it begins the discharge of the battery while the EMS still considers it worth charging it.

These electricity curves are then used to feed the electric equivalent battery model for each scenario. This aging model was presented and used in previous works such as[9], [10] to estimate the battery State of Health (SoH) evolution of batteries according to their use. The End of Life (EoL) of the battery is reached when SOH gets to 64% of the initial battery capacity. The time the SOH needs to reach the EoL results to be the lifespan of the battery in each case. This battery aging model runs on Matlab and considers five aging factors, which are: working temperature, C-rate, DOD, State of Charge (SoC) and time. These five factors are considered to be the ones having major effect on battery aging according to literature [11], [12]. The SoH is defined as the ratio between the actual capacity of the battery against the capacity of the battery when fresh or new.

For the simulation of the battery SoH evolution, the study considered that the battery stays at a constant temperature of 25°C.

## 3 Results and discussion

Fig 3 presents the battery aging model estimation of the SoH evolution along time for each scenario. These results show that, effectively, the battery lifespan using commercial kits is longer than in all the scenarios using an EMS. In fact, its lifespan is more than 3, 5 and 8 times the lifespan of a battery working with an EMS at low, medium, and high use scenarios respectively.

These results may mislead to a conclusion stating that non-intelligent energy management of commercial kits is preferable to extend the battery lifespan. Results would certainly go in the opposite direction if the EMS goal should be oriented to enlarge battery lifespan instead of economic incentives.

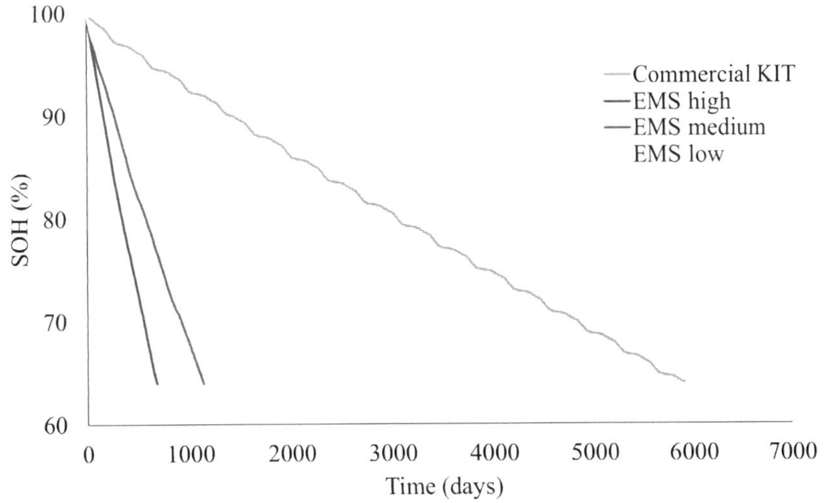


**Fig 3.** SoH evolution and battery lifespan for each scenario

Nonetheless, the reader would have noticed that the relation of the difference in battery lifespan is lower than the one concerning the average cycle use of the battery defined for each scenario. This difference is explained by the fact that batteries age under two circumstances, when they are under use (cycling aging) but also when they are in a stand-by mode (calendar aging).

The aging model used in this study for the SoH estimation considers both aging circumstances (calendar-cycling). In fact, calendar aging is affected by the battery's SoC and temperature during the stand-by lapse. Therefore, for each time-step that the battery is not used there is a loss of capacity that represents an equivalent loss of money in terms of amortization.

The proportion of battery degradation due to calendar aging is presented in Fig 4, showing that the proportion of the battery aging due to stand-by mode in the base case scenario (Kit) represents a 14% of the total aging of the battery. Translated to costs, this means that 521€ are spent just for having the battery ready without any use. In contrast, this factor is much lower for the scenarios that involve an optimized EMS.

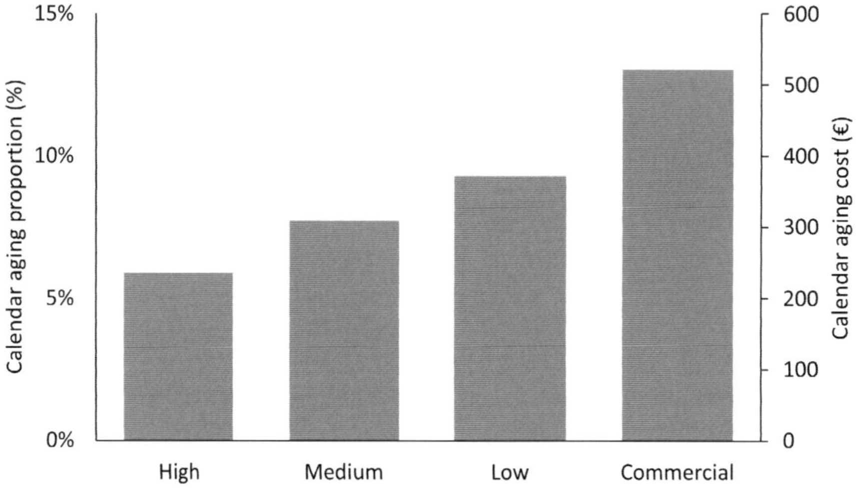


**Fig 4.** Battery degradation proportion due to calendar aging (left) and its attributed cost (right)

This cost analysis can also be related to the quantity of energy delivered by the battery in which case results indicate that, effectively, EMS scenarios exchange more energy than using a commercial solutions. Thus, the cost per kWh delivered by the battery is also lower using smart energy management systems.

Nonetheless, a thorough economic analysis should be carried out, as recent literature indicate that Li-ion batteries for stationary applications are not economically viable due to their high costs [7]. Therefore, it has sense to continue the study regarding possible incentives, energy price scenarios or the use of second life electric vehicle batteries, which have much more competitive costs [13].

## 4 Conclusions

This study analysed the behaviour of several EMS in contrast to commercial kits installed on buildings that count on power generation elements showing that buildings counting on commercial kits tend to underuse the battery capacity.

The study focus the attention on the effect of the differences in use of the battery in relation to its aging. Results show that batteries from buildings that count with an EMS oriented to optimize the electricity bill at the end of the year age faster than those related to commercial kits.

Nonetheless, the analysis indicates that the fact that commercial kits keep the battery in stand-by mode for longer periods of time reverts in a cost of opportunity that EMS manage in a better way.

However, to determine if the EMS strategy is good or not, a more thorough economic analysis should be undertaken considering several tariffs, electricity price evolutions, possible incentives and other alternatives that might decrease battery costs in the nearby future.

The research leading to these results has received funding from the European Union H2020 Programme under GrowSmarter project, Grant agreement no. 646456.

## References

[1] O. Edenhofer, R. Pichs Madruga, and Y. Sokona, *Renewable Energy Sources and Climate Change Mitigation:Special Report of the Intergovernmental Panel on Climate Change* (2012).
[2] B. Scrosati and J. Garche, J. Power Sources **195**, 2419–2430 (2010).
[3] C. Clastres, Energy Policy **39**, 5399–5408 (2011).
[4] H. Beltran, M. Swierczynski, N. Aparicio, E. Belenguer, R. Teodorescu, and P. Rodriguez, *IEEE Int. Symp. Ind. Electron.*, 1604–1609 (2012).
[5] P. Rocha, A. Siddiqui, and M. Stadler, Energy Build. **88**, 203–213 (2015).
[6] A. Barré, B. Deguilhem, S. Grolleau, M. Gérard, F. Suard, and D. Riu, J. Power Sources **241**, 680–689 (2013).
[7] K. Uddin, M. Dubarry, and M. B. Glick, Energy Policy **113**, 342–347 (2018).
[8] L. Igualada and C. Corchero, *IV Congr. Smart Grids*, (2017).
[9] L. Canals Casals, B. Amante García, and M. González Benítez, in *Lect. Notes Manag. Ind. Eng.* 139-151 (2017).
[10] L. Canals Casals and B. Amante García, Batteries **3**(1), 10 (2017).
[11] J. Vetter, P. Nov, M. R. R. Wagner, C. Veit, P. Novák, K.-C. Möller, J. O. Besenhard, M. Winter, M. Wohlfahrt-Mehrens, et al., J. Power Sources **147**, 269–281 (2005).
[12] A. W. Thompson, J. Power Sources **396**, 691–709 (2018).
[13] L. Canals Casals and B. Amante García, J. Green Eng. **6**, 77–98 (2016).

# Towards more resilient electrical networks in urban areas

*Anna* Palau-Mayo[1,*], *Mikel* de Prada[1,*], and *José Luís* Domínguez-García[1,*]

[1]Catalonia Institute for Energy Research, Jardins de les Dones de Negre 1, 2a., 08930 Sant Adrià de Besòs, Barcelona, Spain

**Abstract.** The requirement of system decarbonisation fixed by the EU 2050 plan is leading to an increased establishment of renewable energy sources. Additionally, the emergence of power electronics and ICT technologies has played a decisive role towards a novel distribution electric grid allowing new monitoring, operation and control. In parallel to the energetic transition, an increasing occurrence of extreme weather events and a reinforced concern on climate change leads to the concept of resilience, which is the capacity to adapt and recover from disruptive events in a coordinated procedure. After a fault event, assuming the objective of the system operator is to minimize the load unsupplied, the present study aims at outlining an early research state on the concept of self-healing through the development of a power flow optimization algorithm within a meshed network. Moreover, the effects of integrating Distributed Energy Resources (DER) in order to increase distribution grid resilience as well as to ensure and secure power supply to the system leads to the clusterization of the power system. With controllable technologies, the on-outaged areas are able to disconnect from the main grid, creating islanded microgrids (MGs) which can work autonomously and consequently, increase grid resilience.

## 1 Introduction

The present research is englobed within the project RESCCUE (RESilience to cope with Climate Change in Urban arEas), funded by the European Union's Horizon 2020 Research and Innovation Programme.

Historically, the electrical network has been integrated vertically and centralized, with great generating power plants supplying energy to customers after going across long distances. The current paradigm is forcing the change into a decentralised electrical system due to the appearance of renewable energy sources (RES), energy storage systems (ESS) and electrical vehicles (EV), together with Information and Communication Technologies (ICT). The future energetic model is working towards a distributed energy system with DERs, where the customer will play a central role.

The European Commission has set a target of reaching by 2030 and 2050 a final energy consumption of at least 28% and 55%, respectively, coming from RES [1]. Moreover, the increasing occurrence of natural disasters is shedding doubts to the security and supply

* Corresponding authors: apalau@irec.cat, mdeprada@irec.cat, jldominguez@irec.cat

continuity of the current power system design [2]. According to the United Nations [3], 66% of the population will live in urban areas by 2050 and therefore under natural disasters, an Urban Power System (UPS) with an automatic reconfiguration, or self-healing, of the network would enhance resilience under climate change event and consequently, continuity of supply to customers would be ensured [4]. Various self-healing strategies have been defined for the restoration of a power system assuming the latter is formed by is automated and remotely controlled equipment [5-9]. However, these studies also assume a radial network topology, which evidences the need of developing a strategy for meshed urban networks.

In this paper, an early development of a self-healing strategy on load restoration is presented based on the need of increasing resilience. Ongoing results are being tested on the CIGRE benchmark network at medium voltage (MV) [10].

## 2 Resilience in urban areas

As global warming strikes the world, urban electricity demand is expected to increase from 51% to 58% between 2006 and 2030, which evidences the need of having a resilience strategy management in cities [11]. A coordinated and fast response in front of disruptive events in populated areas leads to the concept of resilience, which is the capability to anticipate, prepare, respond and recover from unexpected extreme events, as observed in Figure 1.

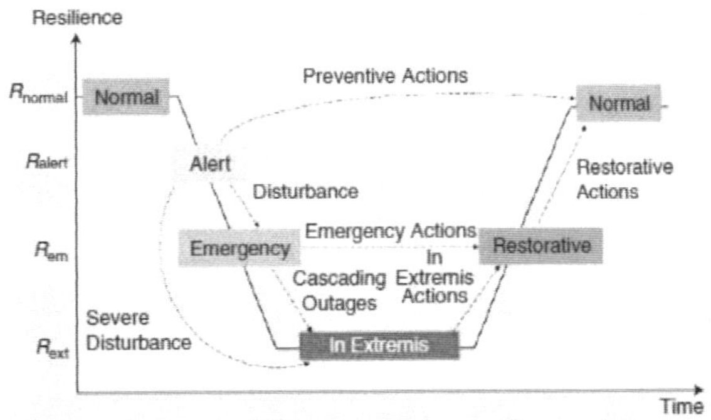


**Fig. 1.** Resilience curve concept [12]

Building a resilient UPS is based on four key factors: meshed operation, integration of DER units, island mode and automation and communication. A meshed operation of a network consists on more than one feeder supplying the load, and therefore continuity of supply is more probable than in radial operation. Moreover, in an UPS where the majority of the lines are underground and hence the reparation of a possible failure is more complex, meshing the network can also be seen as a preventive measure [13]. Continuity of supply is also enhanced by placing DER units in a UPS since spatial diversification of generation reduces the stress of network components and the designed peak power. Moreover, distributed storage units possess the black-start capability necessary for energizing a system immediately after a fault event [14]. Islanding allows for the restoration of that system in case of a major blackout in the main grid, increasing the flexibility of the system. Island mode operation is dependent on the amount of units with a black-start capability in the system, leading to the need of having the aforementioned storage system units to restart the UPS. Regarding the automation and communication, automated and remotely controlled switches would allow a self-healing of the grid instead of a manual reconfiguration which is present nowadays in the majority of UPS when a contingency occurs. Moreover, phase measurement units and local communications become essential in a smart UPS, especially when the system operates as a cluster of microgrids (MGs) [15].

During the planning stage of a power system, static analysis are essential to check the preparedness of the system under unexpected events. To know how resilient the UPS is, contingency analysis are simulated in order to provide information (unsupplied load, violation of network limits, etc.) on possible states of the UPS when a certain equipment is left out of service. In summary, a contingency analysis is essential in determining the current and future potential resilience capacity of an UPS.

## 3 Benchmark network at medium voltage

The CIGRE benchmark at medium voltage (Figure 2) is the network being used for the contingency analysis as it resembles an electric single-line diagram from the perspective of a system operator. Some distributed generation is included in the studied network, as stated in the following paragraph.

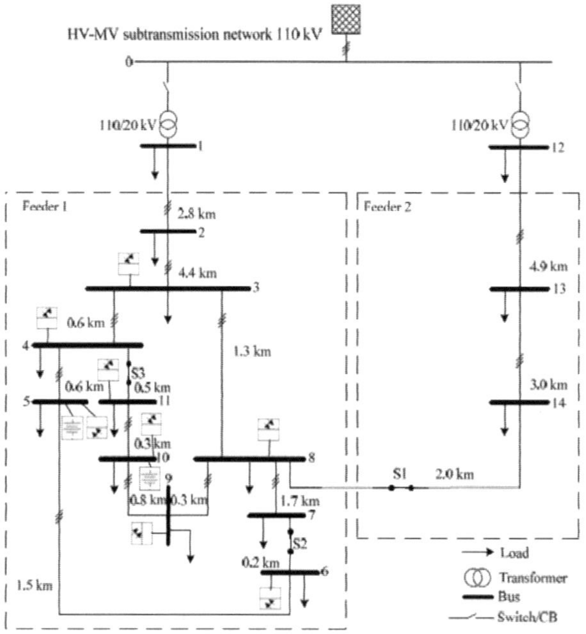


**Fig. 2.** Test case adapted from the CIGRE European network at medium voltage [10]

As it can be observed in Figure 2, there are three switches (S1, S2 and S3) which are either opened or closed depending on the network topology desired. For the present research, all three switches are closed in order to resemble a real meshed UPS. Both 110kV/20kV transformers have a rated power of 25MVA and on-load automatic tap control on the 20kV side, which is limited within the range of 0.9 p.u. to 1.1 p.u. of the nominal voltage. The HV side of the transformer has a no-load tap position control with a voltage ranging between 0.95 p.u. and 1.05 p.u. Regarding MV loads, there are 34.6 MVA, 34.8MVA, 34.0 MVA, 34.5 MVA and 32.0 MVA of peak load (three-phase and in delta connection) at T0, T1, T2, T3 and T4 respectively. It is remarkable to state that loads at Bus 1 and Bus 12 represent 89% of the total peak load since these represent additional feeders. To be able to feed all peak loads simultaneously, the network has been designed with a total capacity of 48MVA. Photovoltaic generation is present in bus 3, 4, 5, 6, 8, 9, 10 and 11 while a battery system is present in both bus 5 and bus 10.

## 4 Self-healing strategy and results

After an unexpected outage in a smart UPS, automated system restoration strategies are a key characteristic in rapidly recovering. Knowing that from a system operator perspective, real-time operational strategies are limited to the amount of sensors and how often these send signals, the proposed algorithm plays with a hypothesis. It is assumed that once an outage happens, the system operator gives the on-outage area of the power system 1 hour to recover and that every 15 minutes, measurements are received. This means that there are a total of five time periods.

For the ongoing research, the self-healing strategy is focused in a two-step strategy: black-start and load restoration Firstly, in the black-start stage, all generators with a black-start capability i.e. being able to turn on without the need of an external supply such as a storage system, must start running in order to re-energize the system in an optimized sequence. Once this first step is fulfilled, the second step of the strategy, which is the point of focus for the present research, is to minimize load unsupplied given certain priorities depending on how critical the infrastructure is. At the same time, for this step, other objectives are the minimization of the line switching actions and the operation of the storage system in order to extend their respective lifecycles. The multi-objective function can is formulated in Equation 1. The binary variable representing the switching between lines has still not been implemented in the current algorithm, so Equation 1 does not reflect this component yet. Let $\alpha$ be the weight parameters, $S_{load_{n,t}}$ be the apparent power of load in node $n$ at time $t$, be $sb$ the base power, be $x_{load_{s,t}}$ the status of the load switch $s$ at time $t$, be $w_{load_{s,t}}$ the load priority at time $t$ and be $x_{bat_{b,t}}^{char}$ and $x_{bat_{b,t}}^{disc}$ the status of the battery switch at time $t$.

$$z = \min\left(\sum_{t\in T}\left(\sum_{s,n\in MAP3}\alpha\frac{S_{\text{load}_{n,t}}}{sb}(1-x_{\text{load}_{s,t}})w_{\text{load}_{s,t}} + \sum_{b\in B}(1-\alpha)(x_{\text{bat}_{b,t}}^{char}+x_{\text{bat}_{b,t}}^{disc})\right)\right) \quad (1)$$

The optimization consists of an optimal power flow algorithm in alternate current (AC) with power system restrictions to comply with, such as line thermal limits, state of charge in the battery and voltage limits. The developed algorithm is to be implemented in the CIGRE network after a failure in the network (e.g. line trip) takes place, in order to see how the network automatically reconfigures to fulfil the main objective: minimization of load unsupplied. The battery is forced to end up with a state of charge equal or above 50%. Loads in bus 1, 5, 8, 9 and 14 are considered critical infrastructure and therefore they are prioritized over the rest of the loads to restore. It is important to clarify that the self-healing algorithm only applies to the on-outage area, but since the current studied network is meshed, the self-healing algorithm applies to the whole power system.

For the present simulation, the AC algorithm has still not been developed. Therefore, the preliminary results in the current paper are achieved after running an optimal power flow with direct current (DC), which means no losses in the system and no reactive power flow. The deterministic algorithm is implemented in GAMS with the *cplex* solver. The fault in the system will be forced between Bus 0 to Bus 1 at T0 and it will be assumed to last during the whole simulation period of 1 hour. In the following Table 1, Table 2 and Table 3, some test results are shown. The pre-fault results refer to the normal-operating conditions while post-fault results refer to the system behaviour once the HV/MV transformer connecting Bus 0 and Bus 1 suffers an outage.

**Table 1.** Active power discharged by batteries (MW)

| | T0 | T1 | T2 | T3 | T4 |
|---|---|---|---|---|---|
| Battery bus 5 | 0 | 0 | 0.6 | 0 | 0 |
| Battery bus 10 | 0.4 | 0 | 0 | 0 | 0 |

(a) pre-fault

| | T0 | T1 | T2 | T3 | T4 |
|---|---|---|---|---|---|
| Battery bus 5 | 0.6 | 0 | 0.6 | 0 | 0.2 |
| Battery bus 10 | 0 | 0.4 | 0 | 0.4 | 0.4 |

(b) post-fault with optimization

**Table 2.** Active power charged by batteries (MW)

| | T0 | T1 | T2 | T3 | T4 |
|---|---|---|---|---|---|
| Battery bus 5 | 0 | 0 | 0 | 0.6 | 0 |
| Battery bus 10 | 0 | 0.4 | 0 | 0 | 0 |

(a) pre-fault

| | T0 | T1 | T2 | T3 | T4 |
|---|---|---|---|---|---|
| Battery bus 5 | 0 | 0.19 | 0 | 0.6 | 0 |
| Battery bus 10 | 0.4 | 0 | 0.4 | 0 | 0 |

(b) post-fault with optimization

**Table 3.** Active power generated by the slack and photovoltaic generators (MW)

| | T0 | T1 | T2 | T3 | T4 |
|---|---|---|---|---|---|
| Slack | 33.2 | 34.3 | 32.2 | 33.3 | 32.1 |
| PV bus 3 | 0.2 | 0.2 | 0.2 | 0.2 | 0.2 |
| PV bus 4 | 0.2 | 0.2 | 0.2 | 0.2 | 0.2 |
| PV bus 5 | 0.3 | 0.3 | 0.3 | 0.3 | 0.3 |
| PV bus 6 | 0.3 | 0.3 | 0.3 | 0.3 | 0.3 |
| PV bus 8 | 0.3 | 0.3 | 0.3 | 0.3 | 0.3 |
| PV bus 9 | 0.3 | 0.3 | 0.3 | 0.3 | 0.3 |
| PV bus 10 | 0.4 | 0.4 | 0.4 | 0.4 | 0.4 |
| PV bus 11 | 0.1 | 0.1 | 0.1 | 0.1 | 0.1 |

(a) pre-fault

| | T0 | T1 | T2 | T3 | T4 |
|---|---|---|---|---|---|
| Slack | 16.1 | 16.2 | 16.1 | 16.7 | 15.1 |
| PV bus 3 | 0.18 | 0.2 | 0.2 | 0.2 | 0.18 |
| PV bus 4 | 0.2 | 0.2 | 0.2 | 0.2 | 0.2 |
| PV bus 5 | 0.3 | 0.3 | 0.3 | 0.3 | 0.3 |
| PV bus 6 | 0.3 | 0.3 | 0.3 | 0.3 | 0.3 |
| PV bus 8 | 0.3 | 0.3 | 0.3 | 0.3 | 0.3 |
| PV bus 9 | 0.3 | 0.3 | 0.3 | 0.3 | 0.3 |
| PV bus 10 | 0.4 | 0.4 | 0.4 | 0.4 | 0.4 |
| PV bus 11 | 0.1 | 0.1 | 0.1 | 0.1 | 0.1 |

(a) post-fault with optimization

During the pre-fault analysis, all load is supplied during all periods of times because there are no contingencies in the network and therefore the capacity of the system is designed to cover up all the demand. On the other hand, during the post-fault simulation, if the optimization described above is implemented, 52%, 52%, 51%, 50% and 53% of the load is supplied at each time period T0, T1, T2, T3 and T4 respectively. Otherwise, if no optimization is implemented after the fault occurs, a much lower percentage of loads would be supplied, more precisely, 5%, 5%, 4%, 4% and 3% respectively. In the following Figure 4, a plot outlining the percentage of supplied load during each of the scenarios is observed.

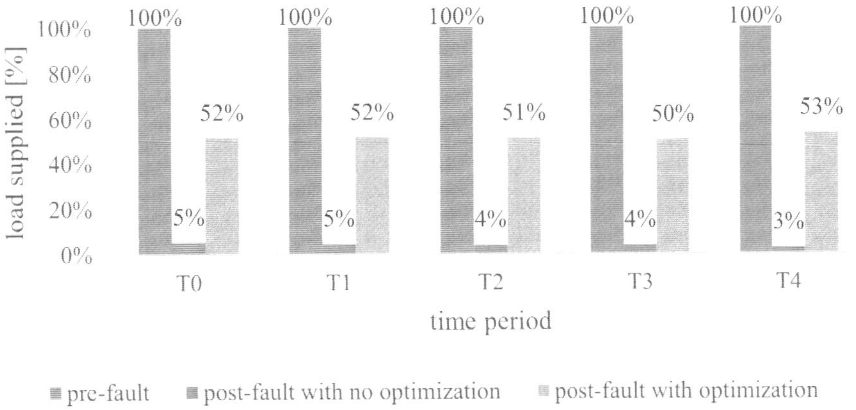


**Fig. 4** Percentage of load supplied in pre-fault and post-fault scenarios [own source]

## 5 Conclusion

The current paper has outlined the need of increasing resilience in a UPS due to the forecasted increase in urban population as well as the factual increase of climate hazard events. The optimization algorithm presented in this paper is at an early research state but even though, results clearly reflect the importance of implementing self-healing techniques under outage events in order to minimize the impact on customers. As future work, an AC optimal power flow in a meshed network is still to be developed. Such algorithm must also include the possibility of clustering the UPS into several smaller power systems in order to decentralise the power generation and increase resilience. Moreover, the optimization algorithm is planned to be scalable to bigger power systems such as a real UPS but the lack of phase measurement units poses a limitation to real life applications.

## References

1. European Commission, "Clean Energy For All Europeans," ed. Brussels: European Commission, 2016.
2. S. Vela, "Identification of Potential Hazards for Urban Strategic Services produced by Extreme Events," public deliverable (RESCCUE Project), 2017.
3. United Nations, Department of Economic and Social Affairs, “World Urbanization Prospects: The 2014 Revision” (ST/ESA/SER.A/352), 2014.
4. T. Huang, "Understanding natural, accidental and malicious threats against secure operation of power systems", ed: Politecnico di Torino - Dipartimento Energia, 2017.
5. C.M. Colson, M.H. Nehrir and R.W. Gunderson, "Distributed multi-agent microgrids: a decentralized approach to resilient power system self-healing," 2011 4th International Symposium on Resilient Control Systems, Boise, ID, 2011, pp. 83-88. doi: 10.1109/ISRCS.2011.6016094
6. A. Golshani, W. Sun, Q. Zhou, Q. P. Zheng and J. Tong, "Two-Stage Adaptive Restoration Decision Support System for a Self-Healing Power Grid," in IEEE Transactions on Industrial Informatics, vol. 13, no. 6, pp. 2802-2812, Dec. 2017. doi: 10.1109/TII.2017.2712147

7. Z. Wang and J. Wang, "Self-Healing Resilient Distribution Systems Based on Sectionalization Into Microgrids," in IEEE Transactions on Power Systems, vol. 30, no. 6, pp. 3139-3149, Nov. 2015. doi: 10.1109/TPWRS.2015.2389753
8. F. Capitanescu, I. Bilibin and J. Sachau, "Towards optimal post-fault self-healing infuture smart distribution grids", in Int. Conf. Renewable Energies and Power Quality (ICREPQ'13), March 20–22, 2013, Bilbao, Spain, 2013.
9. B. Ansari, M. G. Simoes, A. Soroudi and A. Keane, "Restoration strategy in a self-healing distribution network with DG and flexible loads," 2016 IEEE 16th International Conference on Environment and Electrical Engineering (EEEIC), Florence, 2016, pp. 1-5. doi: 10.1109/EEEIC.2016.7555776
10. CIGRE, "Benchmark Systems for Network Integration of Renewable and Distributed Energy Resources," (confidential) 2013.
11. International Energy Agency (IEA), "World Energy Outlook," International Energy Agency, Paris, 2008.
12. M. Panteli and P. Mancarella, "Modeling and Evaluating the Resilience of Critical Electrical Power Infrastructure to Extreme Weather Events," in IEEE Systems Journal, vol. 11, no. 3, pp. 1733-1742, Sept. 2017. doi: 10.1109/JSYST.2015.2389272
13. K. Forssén and K. Mäki, "Resilience of finnish electricity distribution networks against extreme weather conditions," in CIRED Workshop 2016, 2016, pp. 1-4.
14. S. Stout, Hotchkiss, E., "Distributed Energy Generation for Climate Resilience," ed. Minnesota: National Adaptation Forum, National Renewable Energy Laboratory (NREL), 2017.
15. H. Liu, X. Chen, K. Yu and Y. Hou, "The Control and Analysis of Self-Healing Urban Power Grid," in IEEE Transactions on Smart Grid, vol. 3, no. 3, pp. 1119-1129, Sept. 2012. doi: 10.1109/TSG.2011.2167525

E3S Web of Conferences **61**, 00016 (2018) https://doi.org/10.1051/e3sconf/20186100016
*ICREN 2018*

# Design and development of a novel centrifuge ablative pyrolysis approach for biomass conversion to bio-oil and bio-char

*Murlidhar Gupta*[*], *Andrew McFarlan, Leslie Nguyen and Fernando Preto,*

Natural Resources Canada, CanmetENERGY, 1 Haanel Drive, Ottawa, Ontario K1A 1M1, Canada

**Abstract.** Pyrolysis has evolved as a key pre-treatment step to produce renewable fuels and chemicals from agricultural and forestry residues. In the past few years, there have been different directions in the development of pyrolysis reactors. For example, in vortex and cyclone approaches, biomass particles are suspended in a flow of high supersonic velocities to ensure enough centrifugal forces for pressing the particles against the heated reactor surface. Although simple in design, the requirement of large volumes of carrier gases necessitates cumbersome downstream gas separation, resulting in thermodynamic penalties and higher capital equipment costs. In ablative systems, with little or no carrier gases, the key challenge relates to using an appropriate mechanism to continuously apply forces on biomass particles during pyrolysis. In a recent alternative approach, thermo-mechanical rotors at very high rpm have been used to create the required centrifugal forces for pressing the biomass particles against the heated walls of a concentric shell. In the current approach, a modular centrifuge pyrolysis system has been designed using Biot and Thiele numbers as key constraints for characterizing ablative regimes. Unlike other centrifuge pyrolysis reactors, the novel rotor mechanism incorporated in this reactor system facilitates constant centrifugal force as well continuous propagation of biomass feeds. The 10 kg/hr thermo-mechanical pyrolysis system has been successfully commissioned using hardwood sawdust. Properties of bio-oil and bio-char produced in this new reactor have been compared to products from fluid bed pyrolysis system. In addition to its compact and modular design suitable for mobile pyrolysis units, it can be operated in variable regimes of pyrolysis, e.g., slow to fast modes, allowing adjustable product distribution.

## 1 Introduction

Thermochemical conversion of solid biomass residues is considered as a potential pathway for generation of renewable heat and power, as well as value added chemicals. Combustion, gasification and pyrolysis, which vary in degrees of oxidation, are key steps for thermochemical conversion. Unlike combustion processes, pyrolysis is an endothermic

[*] Corresponding author: Murlidhar.Gupta@Canada.ca

thermochemical process and carries some similarity with conventional crude distillation process. In pyrolysis, biomass is heated in absence of any oxidant (air or oxygen) between 400-600 °C to decompose complex biomass molecules, producing solid (bio-char), pyrolysis liquids and gaseous streams (please see Fig.1). These liquids are mixtures of highly oxygenated organic compounds. Although polar and nearly immiscible with conventional crude oil fractions, these pyrolysis liquids are widely referred to as bio-oil because they have the potential to be converted into transportation fuels.

Heat transfer rates play an important role for qualitative and quantitative distribution of solid and liquid yields in pyrolysis processes. Depending upon the heating rates of the biomass particles and the residence time of the vapours, many reactor configurations have been proposed for fast pyrolysis in the last few decades [1,2]. These configurations include shallow moving beds or transported beds (vacuum pyrolysis), auger reactors, fluid beds and ablative reactors.

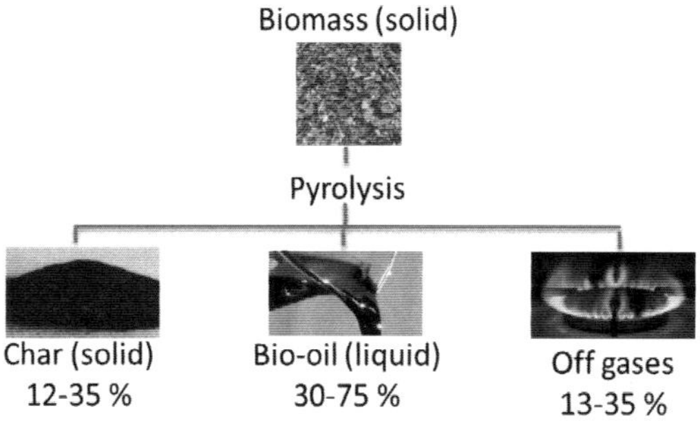


**Fig. 1** Distribution of solid, liquid and gaseous streams in biomass pyrolysis

In moving bed vacuum pyrolysis reactors, the high heating rates are achieved by indirect heat transfer in moving beds through a high temperature molten salt loop and reduced residence time is achieved by pulling out the pyrolysis vapours under vacuum for quick quenching [1,2]. While moving bed pyrolysis processes are capable of handling large particle size distribution and a wide range of feedstocks including plastics and waste tires, the indirect heat transfer and vacuum are the limiting factors in developing a compact moving bed vacuum pyrolysis unit.

In fluid bed fast pyrolysis reactors, the high heat transfer rates are achieved by increasing surface area through reducing biomass feedstocks to very small particle size and transferring the heat through direct exposure to a heat carrier such as hot fluidizing gas (e.g., nitrogen) and heated sand particles. Although some fluid bed pyrolysis reactors have reached demonstration scale, the designs carry inherent requirements of very small particle size (1-3 mm) which increases the milling costs and energy consumption. In addition, the process requires large volumes of carrier gas to be heated, cooled and separated for recycle causing inherent thermodynamic penalties.

In the ablative pyrolysis process, a biomass particle is pressed against a hot surface while the particle moves relative to the heating surface. Due to rapid surface renewal of biomass particles, very high heat transfer rates can be achieved [3]. Since the pyrolysis reaction rates are primarily controlled by the heat transfer mechanisms, the ablative process facilitates rapid pyrolysis of large biomass particles.

In the past few years, there have been different directions in development of ablative reactors. In the vortex and cyclone approaches, biomass particles are suspended in a flow at supersonic velocities to ensure enough centrifugal forces for pressing the particles against the heated reactor surface [1-3]. Although simple in design, the main problem with these reactors is their requirement of large volumes of carrier gas relative to biomass feed which escalates cost due to cumbersome gas separation and resultant thermodynamic penalties [2,4,5].

## 2 Challenges in design and operation of ablative reactor with low or negligible carrier gas

The key challenge relates to using appropriate mechanisms to apply the required force on small and large biomass particles in a continuous ablative pyrolysis reactor, with no or minimal use of a carrier gas. Such ablative reactors would be quite compact and well-suited as mobile pyrolysers, because they process large amounts of biomass with small volumetric footprints.

Bridgewater et al. 2009 used a rotating metal blade to directly press the biomass particles against a heated cylinder [5], whereas, in their pursuit for continuous ablation, Bech et al 2009 used a thermo-mechanical rotor at very high rpm to create centrifugal forces for pressing the biomass particles against the heated walls of a concentric shell [4]. This design used radial blades and spirally twisted flow guide rings welded on the inner surface of the heated reactor wall, for providing continuous centrifugal force and forward motion of the particles. In order to prevent accumulation of unconverted biomass/bio-char inside the reactor, the inner wall of the reactor shell was welded with spirally twisted flow guide rings. Welding of these kinds of guiding rings on the reactor wall is difficult to fabricate and it is difficult to maintain their integrity in a continuous pyrolysis process. If the particle size of the feed changes, the guiding rings need to be changed leading to an expensive and time-consuming process.

The objective of this work has been to design and develop a modular centrifuge ablative pyrolysis reactor, which does not use guiding rings and yet continuously produces pyrolysis product streams comparable to other fast pyrolysis processes, without any biomass or char accumulation in the reactor.

## 3 Materials and Methods

### 3.1 Design approach

Design approach uses dimensionless thermal Biot number, *Bi,* and thermal Thiele number, *Th*. The Biot number is a relative measure of heat flux to particle surface and heat flux to interior of particle and is given by

$$Bi = \frac{hL_0}{\lambda} \tag{1}$$

Where $L_0$ is the characteristic particle length and $\lambda$ is the particle thermal conductivity and $h$ is the external heat transfer coefficient further defined by

$$h = \frac{1}{118(s+1)} d_p \rho g G \tag{2}$$

Here, $d_p$ is the particle diameter, $r$ particle density, $G$ the dimensionless centrifugal acceleration and $g$ is the acceleration due to gravity. For spherical particles, $s$ =2

The thermal Thiele number, *Th,* compares two characteristics times e.g. the heat penetration time and the reaction time and is given by

$$Th = \frac{r_p L_0^2 c_p}{\lambda} \tag{3}$$

Where $c_p$ is the specific heat of biomass particles and $r_p$ is the rate of reaction at an average temperature of 500 °C.

To achieve the ablative regime in the reactor both *Bi* and *Th* were ensured to be greater than one [4,6]. In order to provide continuous propagation and avoid any accumulation of biomass feed and residual char in the shell, a novel rotor design was incorporated into the pyrolysis reactor.

## 3.2 Experimental

A modular ablative centrifuge reactor with a design capacity of 10 kg/hr biomass feed was fabricated and installed on two platforms (See Fig. 2). The upstream section which includes biomass feeder and ablative centrifuge reactor was installed in the upper floor (Mezzanine); see Fig. 2(a). The downstream section, which includes cyclones for solids removal, a direct contact pyrolysis vapours condenser, and gravity separator, was built on the lower platform. (See Fig.2b).

**Fig. 2.** CanmetENERGY's 10 kg/hr pilot scale centrifuge pyrolysis reactor; 2(a) The upstream deck with biomass feeder and reactor; 2(b) The downstream with char separation and vapour handling; 2(c) Compact centrifuge ablative reactor.

The ablative reactor comprises a shell made of a 150 mm (nominal) diameter stainless steel pipe that is externally heated by three band heaters. Band heaters are controlled by temperature monitoring through their respective thermocouples and temperature controllers. The reactor shell houses a rotor whose axis of rotation is fixed by bearing blocks located on either side outside of the reactor housing. The rotor is coupled with a motor that allows the rotor to attain very high rotational speeds up to 10,000 RPM. Control of the motor is performed manually through a variable speed controller. The sealing of the reactor contents from the atmosphere is facilitated by high temperature oil seals within seal assemblies at either end of the reactor. To ensure seals were adequately protected, they were cooled with nitrogen, as well as with water. For hot commissioning experiments, hardwood sawdust was used as the feedstock of choice as the same feed was previously pyrolysed using a 10 kg/hr fluid bed unit with similar balance of plant and this provided an opportunity to compare the products from the two reactor systems.

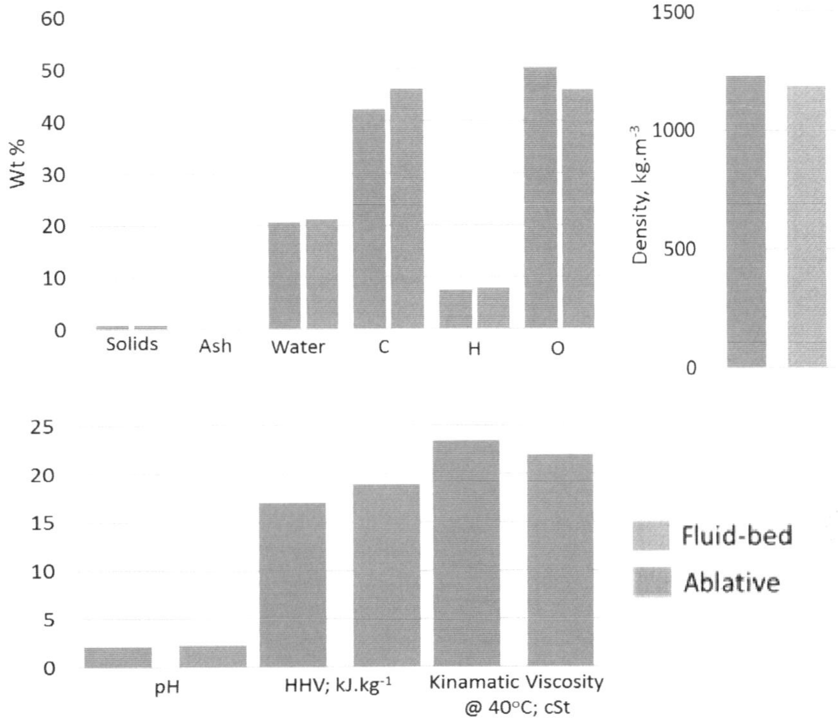


**Fig. 3** Ultimate and proximate analysis of bio-oils obtained from ablative centrifuge fluid bed pyrolysis systems.

## 4 Results and discussion

Prior to the hot pyrolysis runs, fine cellulose powder was used in cold runs to assess the continuity of solid flow at different rotor speeds. The cold tests confirmed the mechanical performance of the reactor design and it attained rotor speeds up to 6500-rpm achieving centrifugal acceleration close to 3.5G. This was achieved without any significant vibrations and without any clogging of flow inside the reactor and cyclone systems.

Subsequently, commissioning pyrolysis runs were carried at 500 °C with hardwood sawdust feedstock. The ablative centrifuge pyrolysis unit ran successfully at a rotor speed of 2000 rpm without any noticeable vibrations and without any clogging of char inside the reactor system. During the hot run operations, the unit was able to maintain fast pyrolysis regime (vapour residence time of less than 2 seconds).

Figure 3 shows the ultimate analysis, density, HHV and kinematic viscosity of bio-oil obtained from the ablative centrifuge pyrolysis unit. These properties were compared with bio-oil produced in a fluid bed unit with the same hardwood sawdust feed and with similar capacity and balance of plant components except the reactor system. The results indicate that although the oxygen content of bio-oil from ablative centrifuge pyrolysis is around 5% higher than fluid bed based system, other properties, e.g., pH, HHV, are nearly the same as for fluid bed reactors. The bio-oil samples from the ablative centrifuge pyrolysis system show approximately 8% less solid loading, which is an indication of less solid attrition in ablative process compared to fluid systems.

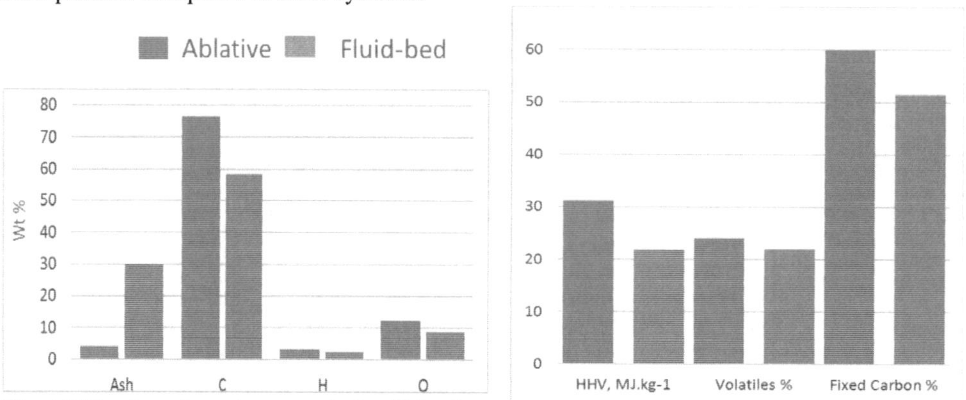


**Fig. 4** Ultimate and proximate analysis of bio-chars obtained from ablative centrifuge and fluid bed pyrolysis systems.

Figure 4 shows the HHV and ultimate analysis of bio-char samples captured at the two cyclones. When compared to bio-char obtained from fluid bed system, results show that ash content in the char from ablative pyrolysis is significantly lower at 4.25% compared to 29.8% in fluid bed system. Similarly, carbon and hydrogen content are also high providing nearly 43% higher heating value for ablative char compared to fluid bed system. Low ash, high fixed carbon and high heating values are desirable in many end use applications that include combustion and metallurgical cokes.

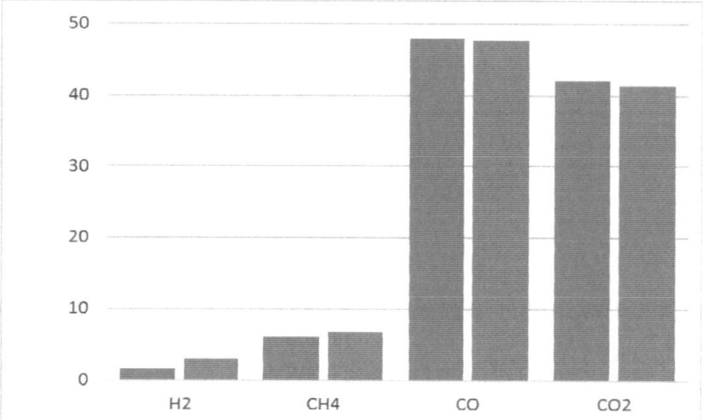


**Fig. 5** Composition of off gases from ablative centrifuge and fluid bed pyrolysis systems.

Figure 5 shows the gas composition of non-condensable gases from ablative pyrolysis of hardwood sawdust on a nitrogen-free basis. When compared with fluid bed off gases, results indicate nearly same concentrations of CO and $CO_2$, which accounts for approximately 90% of total off gases. The mass yield of off-gases for the ablative centrifuge and the fluid bed were also similar at 10.3 and 11.9%, respectively. The concentration of $H_2$ and $CH_4$ from off gas samples in ablative pyrolysis unit are lower than fluid bed system by 47% and 8% respectively, but only account for less than 10% of total gases.

## 5 Conclusions

The cold and hot runs proved that operational continuity of the ablative pyrolysis unit was not hindered by char build-up or clogging inside the reactor. Moreover, the majority of fine char particles exited from the centrifuge reactor to the two cyclones confirming the desired outcome of this novel rotor design.

Bio-oil, char and off gas samples obtained from ablative centrifuge system were compared with the fluid bed pyrolysis system. Results indicate that the product quality in ablative centrifuge system is comparable with other fast pyrolysis systems fed with similar feedstock and having the similar balance of plant components. Future plans include optimisation of unit with low quality forestry and agriculture residues.

Compared to existing state of the art, the novel design of this ablative centrifuge system is extremely simple and compact. Such thermochemical pyrolysers can be operated as distributed systems, under variable pyrolysis regimes ranging from slow to fast pyrolysis, and avoiding large quantities of carrier gas. Although scale up for large centralized pyrolysis plants may be an issue, these ablative centrifuge pyrolysis units can be adapted in a modular approach for large facilities.

Authors appreciate technical assistance from Karam Khosa, Ron Campbell, Sylvain Derome, Mitchell Noy, Benjamin Bronson, and Dillon Mazerolle. Authors acknowledge funding support from Panel of Energy Research and Development, Government of Canada. Authors also acknowledge suggestions from Dr. Peter Gogolek for improving the quality of this manuscript.

## References

1. D. Mohan, C.U. Pittman and P.H. Steele, Energy & Fuels, **20**, 848-889 (2006)
2. D. Vamvuka, International Journal of Energy Research, **35**(10), 835-862 (2010)
3. S. Scholl, H. Klaubert and D. Meier, Science *in Thermal and Chemical Biomass Conversion*, Edited by A.V. Bridgewater and D.G.B. Boocock, **2**, 1372-1379 (2006)
4. N. Bech, M. Boberg Larsen, A.J. Jensen, K. Dam-Johansen, Biomass and Bioenergy **33**, 999-1011 (2009)
5. A.V. Bridgewater, G.V.C. Peacocke and N.M. Robinson, Ablative thermolysis unit", US Patent No. US 7,625,532 B2, (2009).
6. A.V. Bridgwater and D.G.B. Boocock, *Developments in Thermochemical Biomass Conversion*, **1**, ISBN 978-94-009-1559-6 (eBook); DOI 10.1007/978 94-009-1559-6; 1-1648, (1997)

E3S Web of Conferences **61**, 00017 (2018) https://doi.org/10.1051/e3sconf/20186100017
*ICREN 2018*

# A comparative analysis of heat recovery and product stabilization in fluid-bed and ablative pyrolysis systems

*Murlidhar Gupta*[*], *Benjamin King, Fernando Preto and Andrew McFarlan*

Natural Resources Canada, CanmetENERGY, 1 Haanel Drive, Ottawa, Ontario K1A 1M1, Canada

**Abstract.** Nearly all existing commercial pyrolysis technologies employ single-step rapid condensation of vapours from 500 °C to 50 °C using sprays of cold bio-oil or liquid hydrocarbon as a quench fluid. This approach produces raw bio-oil, a non-homogenous mixture of hundreds of oxygenated organic compounds including organic acids and water. Single-step quench also results in loss of high quality heat to the ambient. In this work, a novel 3-stage fractional condensation approach has been proposed. The intent is to produce targeted stable products for value added applications as well to enhance the overall efficiency of pyrolysis processes. The first phase of this research involved modelling and simulation of staged condensation of pyrolysis vapours using Pro/2 process software. A comprehensive pyrolysis model with 13 representative compounds was developed and validated. The Pro/2 model is able to simulate complex condensation of lignin and sugar fractions at high temperatures. Multiple cases involving staged condensation in ablative and fluid bed pyrolysis systems were investigated. In each case, there was a trade-off between high-quality heat recovery and early separation of lignin and sugars from organic acids. Results demonstrated that dew point depression adds additional complexity and limits heat recovery. However, judicious selection of condenser temperatures offers opportunity for early isolation of sugars and lignin from acids, thereby improving product stability.

## 1 Introduction

Pyrolysis is the endothermic decomposition of complex biomass materials at high temperatures in the absence of air or oxygen. Pyrolysis produces solid (bio-char), pyrolysis liquids and gaseous streams. Although polar and nearly immiscible with conventional crude oil fractions, pyrolysis liquids are widely labelled bio-oil as they carry the potential to be converted into transportation fuels e.g. gasoline and diesel etc. [1].

Heat transfer rates play an important role for qualitative and quantitative distribution of solid and liquid yields in pyrolysis processes. Rapid biomass heating and rapid vapour

[*] Corresponding author: Murlidhar.Gupta@Canada.ca

quenching to reduce vapour residence time have been the essence of fast pyrolysis to maximize the yield of bio-oil [2,3].

In last few decades, many new reactor configurations have been proposed for rapid heating of biomass particles in fast pyrolysis [2,3]. These configurations include shallow moving beds or transported beds (vacuum pyrolysis), auger reactors, fluid beds and ablative reactors. While the fluid bed reactor achieve high heat transfer rates by using very small biomass feed size entrained by heated inert gases, the ablative systems use relative motion of biomass feed and a heated surface creating ablation and high heat transfer coefficients [3-5]. Nearly all of the existing pyrolysis technologies employ single step rapid condensation of pyrolysis vapours from 500 °C to 50 °C using sprays of cold bio-oil (or liquid hydrocarbon) as a quench fluid (See Fig. 1).

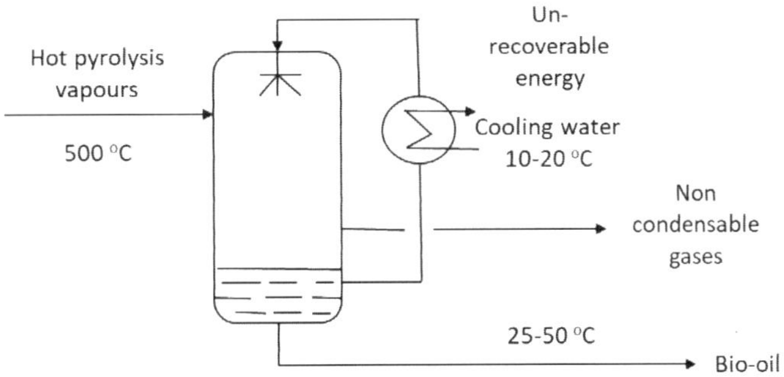


**Fig. 1** Current practice of single step quench of pyrolysis vapours

While single step rapid quench helps maximize the quantity of the liquid product obtained, this liquid product becomes a non-homogenous mixture of hundreds of oxygenated polar compounds and contains large fraction of water and acids. It is widely considered that aging process in bio-oil is initiated by acid-catalysed sugar decomposition forming reactive compounds such as hydroxymethyl furfural. This aging process can be slowed down if acids can be isolated from sugars in the bio-oil [6]. In addition to chemical instability of bio-oil, the conventional one-step quench also leads to loss of high quality heat to the surroundings (primarily cold water utility or ambient air) with no possibility to recover or recycle this energy back to upstream heat requirements for rapid heating.

**Table 1.** Bio-oil representative compounds for Pro/2 simulations

| Functional group | Model compound | Functional group | Model compound |
|---|---|---|---|
| Acids | $C_2H_4O_2$ | Low MW lignin derived compound B | $C_{12}H_8O$ |
| Alcohols | $C_6H_6O_2$ | Extractives | $C_{20}H_{30}O_2$ |
| Ketones | $C_3H_6O_2$ | High MW lignin derived compound A | $C_{16}H_{22}O_4$ |
| Aldehydes | $C_8H_8O_3$ | High MW lignin derived compound B | $C_{10}H_6O_8$ |
| Guaiacols | $C_7H_8O_2$ | Nitrogen compounds | $C_8H_{11}N$ |
| Sugars | $C_{12}H_{22}O_{11}$ | Sulfur compounds | $C_{12}H_8S$ |
| Low MW lignin derived compound A | $C_{14}H_{12}O_2$ | | |

Recently, some researchers have initiated exploring the possibility of staged condensation [7-11]. However, their focus has been to remove water and other low boiling by reducing condenser temperatures in the range of 150 – 4 °C. These low condenser temperatures not only limit any useful heat recovery but also incur more energy loss because of cooling utility requirements for temperatures lower than ambient.

The objective of this study is to develop a preliminary process model for heat recovery and product stabilization in biomass fast pyrolysis processes and simulate multistage condensation and product fractionation for ablative and fluid-bed pyrolysis systems. The intent is to separate lignin and sugars from acids to enhance product stability and quality produce targeted stable products for value added applications as well enhance the overall efficiency of pyrolysis processes.

## 2 Materials and methods

Pro/2 simulation software (Schneider Electric®) was used to develop the pyrolysis model. Pro/2 is also capable of modelling solids, a desired feature for incorporating biomass and char in simulations. Since biomass, char and bio-oil are

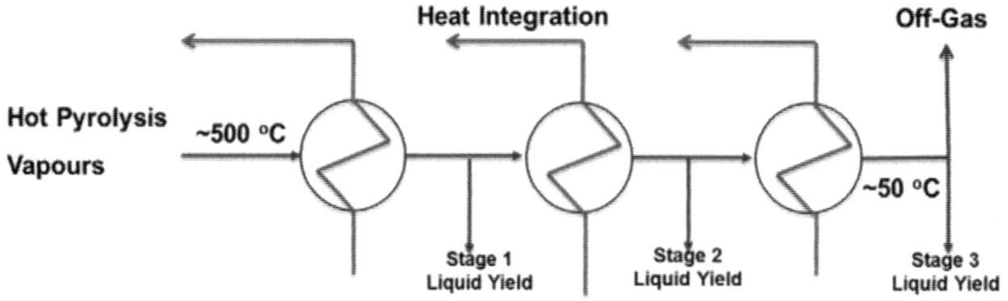


**Fig.2.** 3-stage fractional condensation concept for heat recovery and product stabilization.

complex materials and are not defined in Pro/2's chemical library, it was necessary to identify representative compounds in Pro/2's chemical library which could simulate key functional compounds in bio-oil. Our laboratory (CanmetENERGY-Ottawa) had previously

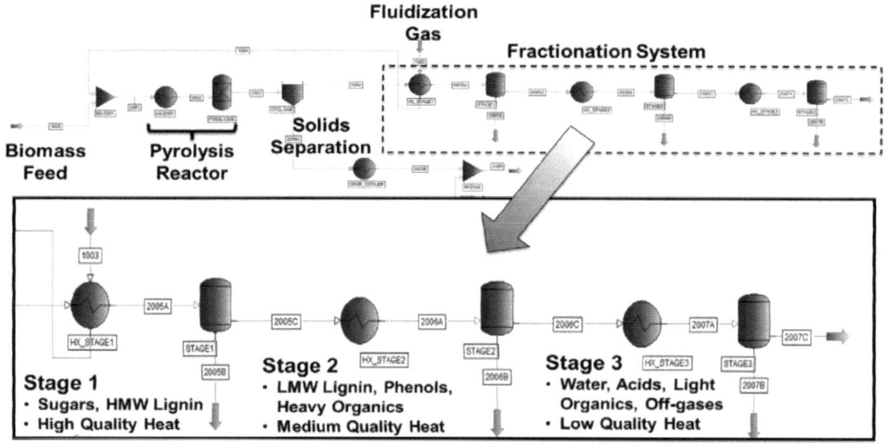


**Fig. 3**. A Pro/2 process model for biomass pyrolysis with three-stage condensation.

developed a Pro/2 process model using 10 representative compounds [12]. The previous pyrolysis model used single step quenching and considered bio-oil as a simple fluid. Model compounds considered in previous model, did not adequately represent the physical properties or complexity of bio-oil condensation. For example, in the previous model, ethyl phenol which represented phenolic monomers, was the compound with highest melting point (46 °C). Whereas in the literature, common compounds in bio-oil such as pyrolytic lignin (approximately 20 wt. % of bio-oil) and sugars (approximately 30 wt. % of bio-oil) have been reported to have a melting point in the range of 180 – 220 °C [1,13]. Lack of adequate compounds representing sugars and lignin, in previous model was found to be unsuitable for simulating condensation of high melting sugars and lignin in multistage quenching.

Subsequently, a pyrolysis model with 13 representative compounds was developed. For the purpose of simulating condensation process, representative compounds were chosen from Pro/2 library to have as close as possible molecular structures as well the phase behaviour (melting and boiling points) as the key functional groups in bio-oil as shown in Table 1.

## 3 Results and discussions

The three-stage condensation concept (as shown in Fig. 2) targeted to bring down pyrolysis temperatures from 500 to 50 °C offering three different qualities of heat available for recovery. Fig. 3 shows sample Pro/2 process model used for simulating this three stage concept.

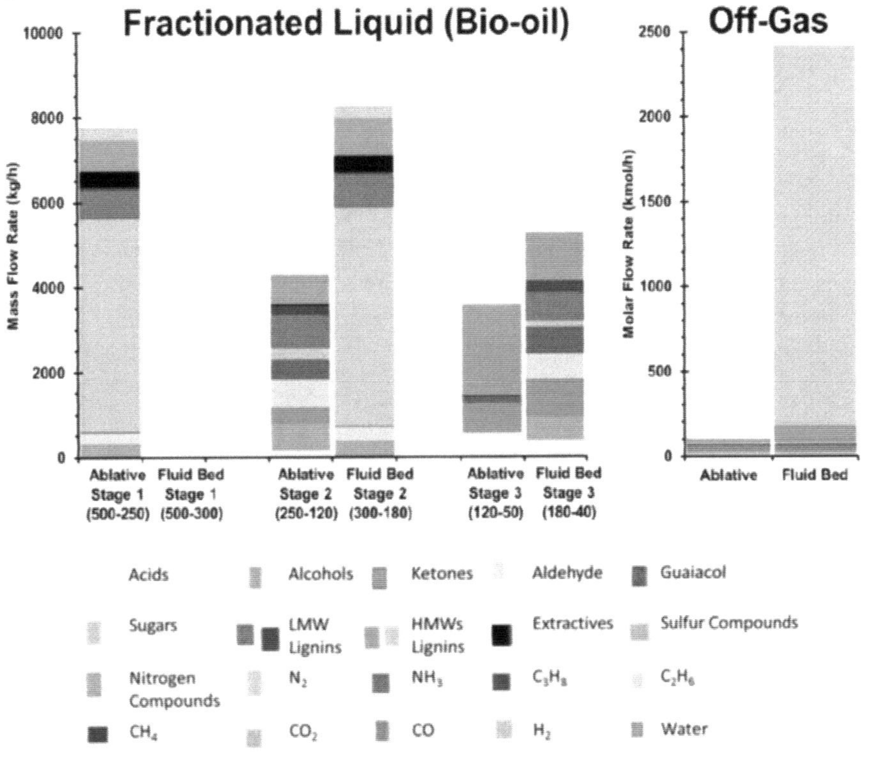


**Fig.4.** A comparative analysis of 3-staged condensation in ablative and fluid bed systems

The revamped model with representative compounds shown in Table 1 was able to effectively simulate lignin and sugar condensation. Prior to conducting simulations on pyrolysis vapour condensations, model results were verified with NREL's 2013 model [14].

Multiple cases involving staged condensations in ablative and fluid bed pyrolysis systems were investigated. These two system represented low and high carrier gas to biomass ratios, respectively. In each case, the intent was to reach a trade off between high quality heat recovery and early separation of lignin and sugars with that of acids to enhance product stability and quality. Fig. 4 shows striking differences between ablative and fluid bed systems. For example, high amounts of fluidization gas causes dew point depression in the product vapours. As a result, pyrolysis vapours containing HMW lignin, LMW lignin and extractives that typically exit the system as liquid in Stage 1 of an ablative process, remain vapours in fluid bed process. At around 250 °C, almost (95%) of high melting sugar and HMW lignin as well extractives are condensed early in the first stage while most of the acids and water are discarded in the third stage. As lignin and sugar are very distinctive in chemical properties and in term of energy density (33 MJ/kg and 17 MJ/kg respectively – See Fig. 5a), their early recovery and isolation from acid catalyzing reactions offers additional opportunities for their immediate applications or for their further selective upgrading.

In the fluid bed system, in the first stage, no condensation occurs when the condenser temperature is around 300 °C. This is disappointing from fractionation perspective. However, this also negates any change of fluid state in first stage, thus offering more flexibility for fouling-free heat recovery. The effect of dew point depression can be further noticed in stage 3 and in off gas.

Compared to ablative pyrolysis, large fractions of water and acid vapours slip through in the off gases, which also contain significant amount of inert fluidization gases. High relative

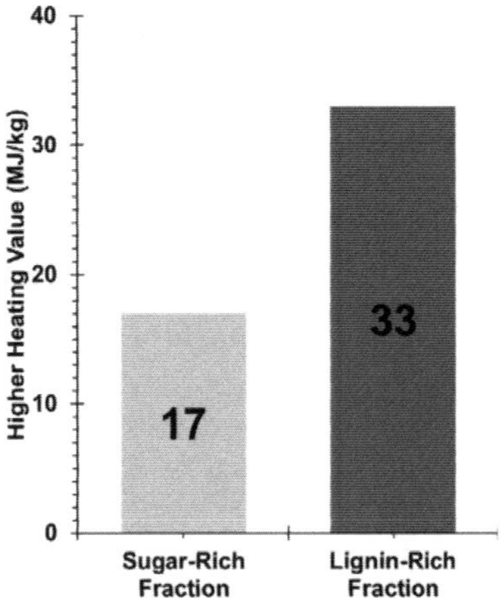


**Fig.5a.** Relative higher heating values of sugar rich and lignin rich fractions in stage 1 of ablative pyrolysis unit.

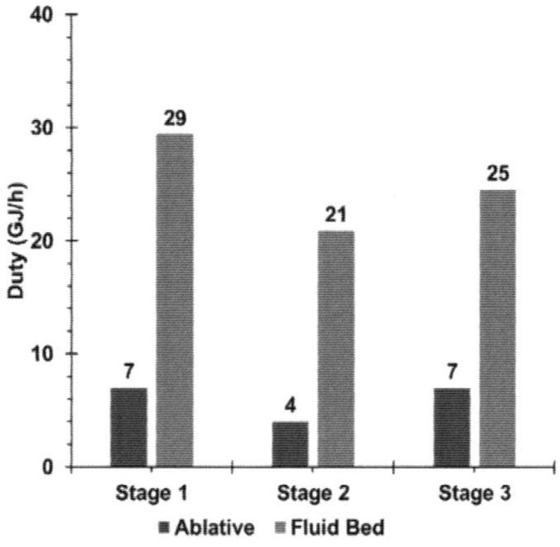


**Fig. 5b.** Relative recoverable heat available for 3 staged vapour condensation in ablative and fluid bed systems.

volume of fluid bed gases could be a matter of concern as any gas recycle strategy will require larger equipment and larger plant footprints. Thus, need for heat recovery in fluid bed is much more intense than in ablative systems (See Fig. 5b).

## 4 Conclusions

This work provided first of a kind simulation study on advanced pyrolysis processes with a focus on heat recovery and product stabilization. In this work, a Pro/2 pyrolysis model with 13 representative compounds was used to model bio-oil. This model was used to compare fluid bed and ablative pyrolysis systems. The outcome demonstrated that staged condensation and heat recovery offer better product selectivity and better process efficiency than single-step pyrolysis vapour condensation. Judicious selection of condenser temperature configurations offers opportunity for early isolation of sugars and lignins from acids, thus reducing instability of product stream. Quality of product fraction and heat recovery are dependent on type of reactor. Given the high ratios of fluidization gas to biomass ratio in fluid bed pyrolysis, heat recovery and recycle are very crucial for these systems.

The financial support from Natural Resources Canada, the Program of Energy R&D is greatly acknowledged.

## References

1. D. Mohan, C.U. Pittman and P.H. Steele, Energy & Fuels, **20**, 848-889 (2006).
2. D. Vamvuka, International Journal of Energy Research, **35**(10), 835-862 (2010)
3. R.H. Venderbosch and W. Prins, Biofuels, Bioprod. Bioref., **4**, 178-208 (2010)
4. S. Scholl, H. Klaubert and D. Meier, Science *in Thermal and Chemical Biomass Conversion*, Edited by A.V. Bridgewater and D.G.B. Boocock, **2**, 1372-1379 (2006)
5. N. Bech, M. Boberg Larsen, A.J. Jensen, K. Dam-Johansen, Biomass and Bioenergy **33**, 999-1011 (2009)
6. J. Meng, A. Moore, D. Tilotta, S. Kelley and S. Park, ACS Sustainable Chem. Eng., **2**, 2011-2018 (2014)
7. A. Oasmaa,, I. Fonts, M-R. Pelaez-Samaniego, M-E. Garcia-Perez, M. Garcia-Perez, Energy & Fuels, **30**, 6179-6200 (2016)
8. C. Lindfors, E. Kuoppala, A. Oasmaa, Y. Solantausta, V. Arpiainen, Energy & Fuels, **28**, 5785–5791(2014)
9. R.J.M. Westerhof, D. W.F. Birlman, M. Garcia-Perez, Z. Wang, S. R. G. Oudenhoven, W.P.M. Swaaij and S. R. A. Kersten, Energy and Fuels, **25**, 1817-1829 (2011)
10. T. Schulzke, S. Conard and J. Westermeyer, Biomass and Bioenergy, **95**, 287-295 (2016)
11. A.S. Pollard, M.R. Rover and R.C. Brown, J. Analytical and Applied Pyrolysis, **93**, 129-138 (2012)
12. M. Gupta, R. Pigeon, A. McFarlan and C. Fairbridge, *tcbiomass2013: The International Conference on Thermochemical Conversion Science*, 3-6 September 2013, Chicago, IL, USA, Available at www.gastechnology.org/tcbiomass/Pages/2013-presentations.aspx
13. I. Brodin, *Chemical Properties and Thermal Behaviour of Kraft Lignins, PhD Thesis,* KTH Royal Institute of Technology, Stockholm, 1-47 (2009). Available at www.diva-portal.org/smash/get/diva2:234300/fulltext01
14. S. Jones, P. Meyer, L. Snowden-Swan, E. Tan, A. Dutta, J. Jacobson, K. Cafferty, *U.S. Department of Energy*, PNNL-23053, 1-86 (2013); www.nrel.gov/docs/fy14osti/61178.pdf

# Partial deoxygenation of biomass derived pyrolysis liquids

*Murlidhar Gupta*[*], *Jacques Monnier*, *Eric Turriff*, *Mark Boyd*

Natural Resources Canada, CanmetENERGY, 1 Haanel Drive, Ottawa, Ontario K1A 1M1, Canada

**Abstract.** Biomass pyrolysis liquids (also known as bio-oil), are derived from renewable lignocellulosic biomass residues by fast pyrolysis process. These second-generation oxygenated hydrocarbon resources have the potential to partially substitute for petroleum-derived feedstocks and thus enhance the economic and environmental sustainability of our natural resources. However, in contrast to petroleum fuels, biomass-derived pyrolysis liquids contain a large amount of oxygen, usually 40-50% wt% (wet basis). This undesirable high oxygen content in pyrolysis liquids is considered as the primary reason for its high polarity, high acidity, lower stability, lower energy density and very low miscibility with conventional crude refining feedstocks. There are two major pathways for upgrading the pyrolysis liquids. While hydrodeoxygenation route is one of the most explored options, it requires production and supply of large amounts of expensive hydrogen at high pressures, mandating large and centralized upgrading plants, and thus large capital investment. In this paper, we discuss an alternative method of pyrolysis liquid upgrading, using cheap and affordable hydrogen donor additives and catalysts to promote partial deoxygenation at near atmospheric pressure. This approach is preferably to be used as a pre-treatment and stabilizing method for pyrolysis liquids in the close vicinity of remote biomass pyrolysis plants. The pre-treated oil, then can be shipped for further hydrocracking process in a centralized co-processing facility. Preliminary results from the initial proof of concept experiments involving a 200 g/h gas-phase continuous fast catalytic cracking system with continuous coke removal to enhance deoxygenation performance are presented. These results indicate positive impact of catalyst bed on quality and yield of the upgraded bio-oil product in terms of pH, viscosity, degree of deoxygenation, oil yield and concentration of hydrogen in the off gases.

## 1 Introduction

Thermochemical conversion of agricultural and forestry residues is considered a potential pathway for generation of renewable heat and power as well as value added chemicals. Unlike petroleum resources, which are found in a highly concentrated form in nature, biomass residues are highly heterogeneous and dispersed along irregular landscapes. The

[*] Corresponding author: Murlidhar.Gupta@Canada.ca

spatial dispersion of raw biomass requires high fractions of parasitic loss per unit of energy during its collection and transportation (haul) to central processing sites such as refineries [1].

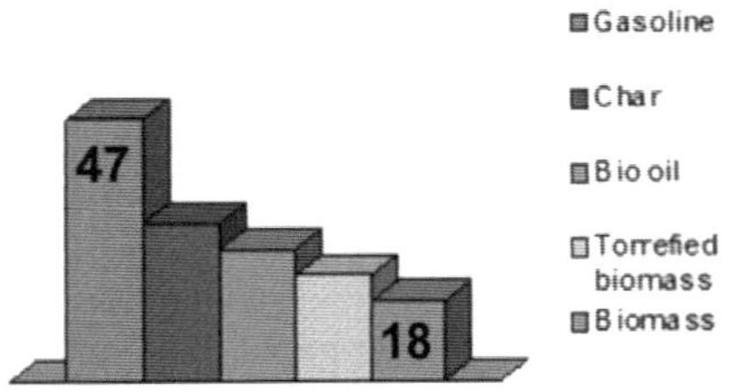


**Fig. 1.** Relative energy density of bio-oils compared to raw biomass and gasoline.

One of the key pathways to convert the raw biomass into higher density intermediates is fast pyrolysis process. Unlike combustion and gasification processes, pyrolysis is an endothermic thermochemical process and carries some similarity with conventional crude distillation process. In pyrolysis, biomass is heated in absence of any oxidant (air or oxygen) between 400-600°C to decompose complex biomass molecules, producing solid (bio-char), pyrolysis liquids and gaseous streams. These liquids are a mixture of highly oxygenated organic compounds. Although polar and nearly immiscible with conventional crude oil fractions, these pyrolysis liquids are widely called bio-oil as they carry the potential to be converted into transportation fuels e.g. gasoline and diesel, etc. [2]. Major issues with bio-oils are their low pH, increase in viscosity with time (e.g., aging) and close to 50% oxygen content resulting in low energy density (see Fig. 1).

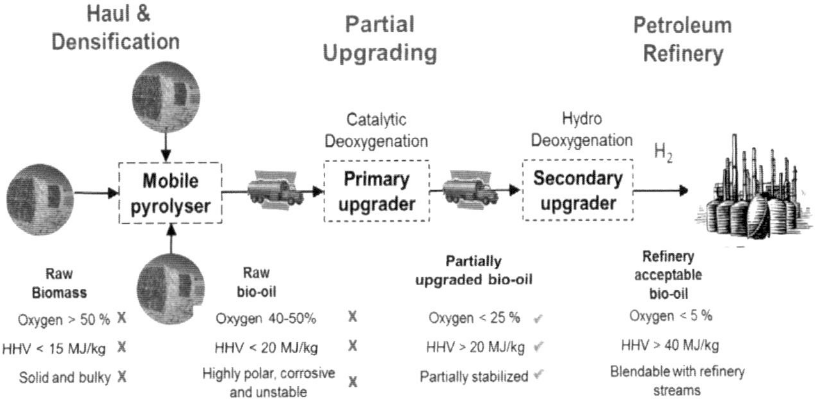


**Fig. 2.** Sequential upgrading and hauling of bio-oil using partial deoxygenation

### 1.1 Bio-oil upgrading

In order to enhance bio-oil, the pH, stability and energy density, oxygen content of the bio-oil needs to be addressed. In line with hydrodesulphurization (HDS) process from the petroleum refinery industry, hydrogen pressures greater than 75 bar have been widely suggested for nearly complete hydrodeoxygenation (HDO) of bio-oil [3]. However, producing and handling high pressure hydrogen is technically challenging and expensive at remote sites where most of the agricultural and forestry residues are naturally available.

https://doi.org/10.1051/e3sconf/20186100018

Instead of using a highly sophisticated and difficult to maintain HDO facility at a remote sites, an alternate option is primary upgrading through in-situ low pressure partial deoxygenation of raw bio-oil in close vicinity of a biomass pyrolysis facility without using an external hydrogen supply. This way partially reduced levels of oxygen results in higher pH, better stability and greater energy density, making it suitable for long-term storage and transportation to centralized upgrading sites where appropriate resources for secondary upgrading of this partially deoxygenated bio-oil may be available. Fig. 2 shows a conceptual schematic of sequential upgrading and hauling of biomass residues to a petroleum refinery where the deoxygenated bio-oil can be further upgraded with existing refinery infrastructure.

The objective of this study is to develop a simple and partial deoxygenation process using suitable diluents and catalysts, which can be easily integrated at remote sites.

## 2 Materials and methods

Bio-oil sourced from forestry residues were blended with two diluents. The purpose of using these diluents was to prevent coking in the feedline and to explore the possibility of them acting as hydrogen donor. Diluent "A" was a non-polar refinery fraction and was used to make a stable feed blend - for continuous feeding of bio-oil into the reactor system. Diluent "B" was a polar low value industrial by-product. An auger based reactor with a feed capacity of 200 g.hr$^{-1}$ was used to quickly vapourize bio-oil and diluent mixtures and to ensure quick removal of char from vapours proceeding to the fast catalytic cracking reactor (FCCR) (See Fig. 3). The FCCR operated between 400-500°C and near-atmospheric pressure, with a packed bed of metal based catalysts. Vapours downstream of reactor were subjected to primary quench at 5°C followed by a secondary quench at -15°C to capture condensates for characterization. Off gases were measured and sampled prior to directing them to vent.

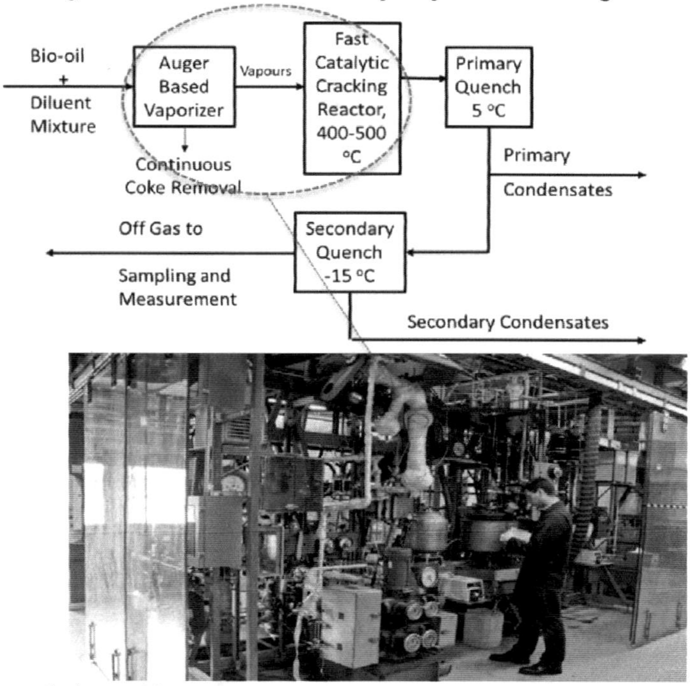


**Fig. 3.** Fast catalytic cracking reactor

## 3 Results and discussion

During all experiments, successful mass balances (>95%) were achieved. Fig. 4 shows significant kinematic viscosity decrease from 38 cSt to nearly 3 cST for feed (with diluent A) and upgraded bio-oil (product) respectively. However when metal based catalyst is used this decrease diminishes from 38 cSt for feed (with diluent A) to nearly 5 cSt for product stream. This shows that use of metal catalyst under the given conditions did not make a difference.

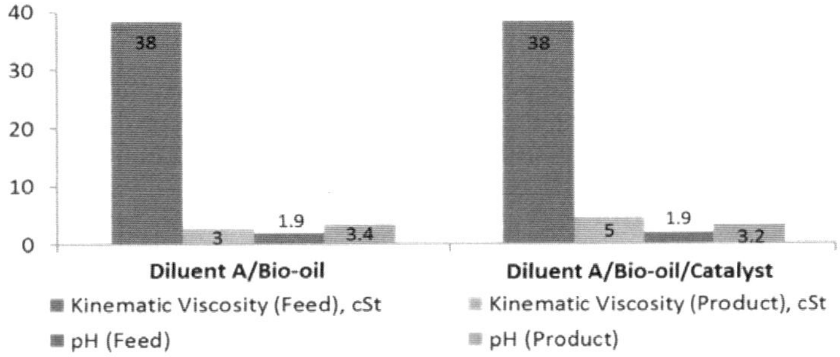


**Fig. 4.** Comparative performance of bio-oil feed with and without metal based catalyst.

pH measurement was used to determine the acidic characteristic of bio-oil and of the product. Fig.4 shows the acidity of the product decreases with pH increases in the product when compared to feed. This is a desirable outcome. However, the effect of catalyst on pH is not clear, as with or without catalyst the increase in pH is around 68%.

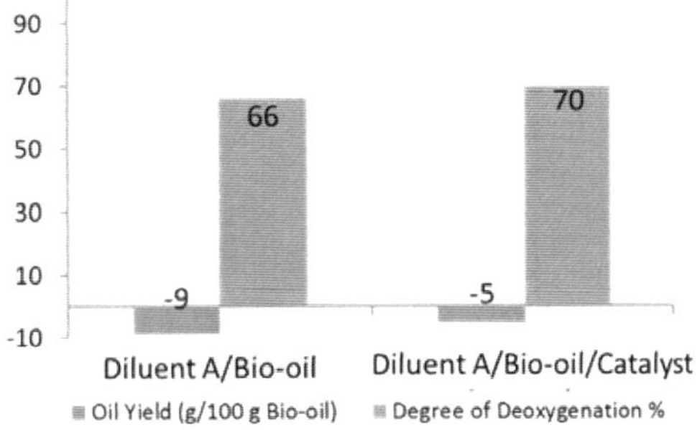


**Fig. 5.** Oil yield and degree of deoxygenation

A positive oil yield shows that the process has not consumed the diluent, rather the bio-oil is being consumed to form the product. This is a desired outcome e.g. diluent fraction can be recycled into the process whereas a negative yield would have indicated that diluent had been consumed in the process and the process would require additional make up for diluent. In case of Diluent A/Bio-oil feed with and without catalyst, we noticed a negative yield ranging from -5 to -9 g per 100g bio-oil, indicating that process with this outcome is not acceptable (See Fig. 5).

The mean value of degree of deoxygenation is more than 50% for the cases with and without catalyst (see Fig. 6). However, the degree of deoxygenation should always be analysed in conjunction with the oil yield. For a meaningful upgrading of bio-oil, both values should be positive and the yield needs to be enhanced through proper process optimization of parameters and requires further investigation.

Fig. 6 shows the hydrogen concentration in the off gases. In both cases e.g. with Diluent A and with Diluent B, use of catalyst shows significant increase in hydrogen concentration in the off gases- from 8 to 21 % for diluent A and to 17% for diluent B. This is an interesting outcome although not desired for producing liquid fuels but highly desirable for producing hydrogen rich gases for fuels and feedstock.

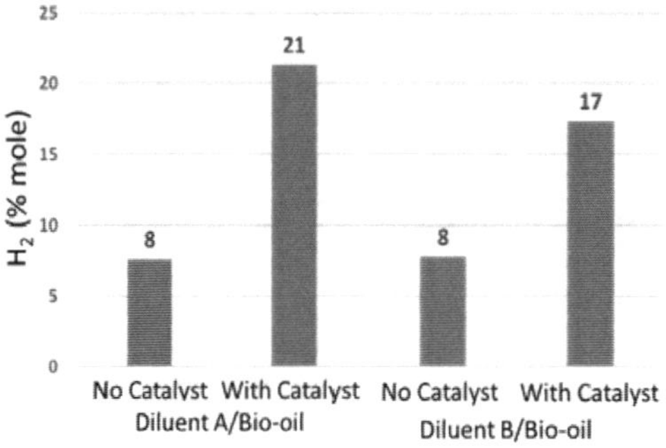


**Fig. 6.** Concentration of hydrogen in the off gases.

## 4 Preliminary conclusions

- Significant reduction in kinematic viscosity was observed.
- pH of product nearly doubled, indicating lower corrosiveness.
- Degree of deoxygenation achieved was greater than 50%, however negative values of yield were also observed.
- Significant increase in the hydrogen concentration was observed in the off gases using catalytic cracking.
- Future efforts should be focussed on optimizing the process with better configuration of reactor temperature, diluents and catalysts.

The financial support from Natural Resources Canada, the Program of Energy Research and Development, is acknowledged. Technical help from Tim Shaw is greatly appreciated.

## References

1. M. Vaezi, A.K. Katta and A. Kumar, Biosystems Engineering **118**, 52-67 (2014)
2. D. Mohan, C.U. Pittman and P.H. Steele, Energy & Fuels, **20**, 848-889 (2006)
3. P.M. Mortensen, J.-D. Grunwaldt, P.A. Jensen, K.G. Knudsen and A.D. Jensen, Applied Catalysis A: General, **407**, 1-19 (2011)

# AUTHOR INDEX